The Mathematical Imagination

The Mathematical Imagination

On the Origins and Promise of Critical Theory

Matthew Handelman

FORDHAM UNIVERSITY PRESS

NEW YORK 2019

This book is freely available in an open access edition thanks to TOME (Toward an Open Monograph Ecosystem)—a collaboration of the Association of American Universities, the Association of University Presses, and the Association of Research Libraries—and the generous support of Michigan State University. Learn more at the TOME website, available at: openmonographs.org.

Through the generous funding of Michigan State University, this publication is available on an open access basis from the publisher's website.

Fordham University Press gratefully acknowledges financial assistance and support provided for the publication of this book by Michigan State University.

Visit us online at www.fordhampress.com.

Library of Congress Cataloging-in-Publication Data available online at https://catalog.loc.gov.

Printed in the United States of America

21 20 19   5 4 3 2 1

First edition

CONTENTS

Introduction: The Problem of Mathematics in
Critical Theory   1

1. The Trouble with Logical Positivism: Max Horkheimer,
    Theodor W. Adorno, and the Origins of Critical Theory   25

2. The Philosophy of Mathematics: Privation and
    Representation in Gershom Scholem's Negative Aesthetics   65

3. Infinitesimal Calculus: Subjectivity, Motion, and
    Franz Rosenzweig's Messianism   104

4. Geometry: Projection and Space in Siegfried
    Kracauer's Aesthetics of Theory   145

    Conclusion: Who's Afraid of Mathematics? Critical
    Theory in the Digital Age   187

    Acknowledgments   201

    Notes   205

    Bibliography   245

    Index   269

The Mathematical Imagination

The Problem of Mathematics in Critical Theory

Humanists are learning mathematics—again.[1] Amidst a renewed sense of crisis in literary, cultural, and language studies, many humanists have turned to mathematics and digital technologies based on mathematical processes in hopes of modernizing and reinvigorating humanistic inquiry. Literary, cultural studies, and media studies scholars as well as historians are using algorithms to read novels, making digital maps to plot the geographies of films, using online tools to annotate and publish texts collaboratively, and applying other computational technologies to explore historical and literary records. According to proponents of such new methods, the so-called digital humanities promise to bring the analytic power of computation to bear on the study of culture and the arts, lending the humanities a more public face and, thus, renewed relevance in the early twenty-first century.

Of course, not everyone shares the digital humanists' enthusiasm and optimism. One recent op-ed in *The Atlantic* alleges that the digital turn in the humanities simply reacts to economic worries about funding increases in and administrative emphasis on STEM fields (science, technology, engineering, and mathematics).[2] The proposed digital rejuvenation of the humanities threatens to forfeit precisely what the critical study of art, literature, and history offer our advanced scientific society: access to concepts such as understanding and empathy that, by their very nature, resist quantification. Indeed, other critics of the digital humanities worry that, beyond not bringing anything essentially new to humanistic inquiry, the climate around the digital, in fact, eschews the rigorous historical research and critical discourse central to the humanities.[3] If the digital humanities embrace the tech industry, do the humanities not also acquiesce to the merger of technology and industry, whose mechanisms of manipulation and control critical theory seeks to expose and oppose?

What often goes unacknowledged in these contemporary debates is the long history of similar disagreements over epistemology that date back to the very inception of critical theory. As Max Horkheimer (1895–1973) and Theodor W. Adorno (1903–1969) first conceived of it in the 1930s, critical theory steadfastly opposed the mathematization and quantification of thought. For them, the equation of mathematics with thinking, embraced by their intellectual rivals, the logical positivists, provided the epistemological conditions leading reason back into the barbarism and violence that culminated in World War II and the Holocaust. However, the fact that Horkheimer and Adorno interwove mathematics with the dialectics and downfall of enlightenment obscures how mathematics provided some of their intellectual forerunners and friends—Gershom Scholem (1897–1982), Franz Rosenzweig (1886–1929), and Siegfried Kracauer (1889–1966)—with concepts, metaphors, and tools that helped negotiate the crises of modernity. Although Scholem, Rosenzweig, and Kracauer are not often counted as critical theorists, we can find in their work the potential for theory that is at once mathematical and critical.[4] In particular, their theories of aesthetics, messianism, and cultural critique borrow ideas from mathematical logic, infinitesimal calculus, and geometry to theorize art and culture in ways that strive to reveal and, potentially, counter the contradictions of modern society. By revisiting and rethinking the origins of critical theory, this book seeks to recapture the potential contribution that mathematics holds for the critical project. To understand the influence of mathematics on Scholem, Rosenzweig, and Kracauer is to uncover a more capacious vision of critical theory, one with tools that can help us confront and intervene in our digital and increasingly mathematical present.

The Eclipse of Mathematics in Critical Theory

In 1935, Edmund Husserl saw the world of reason that he had helped construct crumbling before him. A founder of the philosophical school of phenomenology earlier in the century, Husserl held a series of lectures that year in Prague recounting how, over time, the positivistic special sciences had eliminated all the genuine problems of reason—the question of rational knowledge, the ethics of truly good action, and the notion of values as values of reason. At some point, Europeans had traded a mode of thinking genuinely concerned with reason, ethics, and values—the basic questions of humanity and their meaning in life—for the facts of science and the formulae of mathematics. First published in Belgrade as "Die Krisis der europäischen Wissenschaften und die transzendentale Phänomenologie" ("The Crisis of the European Sciences and Transcendental Phenomenology," 1936), these lectures took on a very different tone than Husserl's other introductions to phenomenology, not least because they could not be delivered or published in Nazi Germany (Husserl was of Jewish descent).[5] Instead of the "Age of Enlightenment" producing the great philosophers to whom Husserl had turned in *Cartesian Meditations* (1931), it now appeared as if the advent of the mathematical natural sciences in the Enlightenment had been the progenitor of a radical turn away from reason in philosophy, manifest in a new type of thought that threatened to "succumb to skepticism, irrationalism, and mysticism."[6] For Husserl, stripped of his German citizenship and removed from the roster at the University of Freiburg, the ramifications of the situation were undeniable. This was not merely a crisis in the natural sciences or philosophy but a fundamental problem with knowledge and reason as such, as implied by the broader German term *Wissenschaft* (literally, body or collection of knowledge). And yet Husserl thought crisis could still be avoided and Europe could still be saved, but only if, as he put it in the preface to the 1936 publication, the heirs of the Enlightenment embraced "the unavoidable necessity of a transcendental-phenomenological re-orientation of philosophy."[7] Husserl died in April 1938; a year later, Germany invaded Prague on its way to total war.

Europe and its sciences had, of course, been in crisis for decades. A "crisis of language" (*Sprachkrise*) had plagued the intellectual life of fin-de-siècle Vienna, inspiring the work of poets such as Hugo von Hofmannsthal, cultural critics such as Fritz Mauthner, and philosophers such as Ludwig Wittgenstein. For these thinkers, language no longer offered a reliable means of capturing and communicating experience and thought, the more problematic aspects of which Wittgenstein famously recommended that we pass over in silence.[8] In 1922, the idea that history called into question the state, morals, and religion, instead of providing their justification, signaled to Ernst Troeltsch a "crisis of historicism" (*Krise des Historismus*).[9] For Troeltsch and others, the idea that there might be no moral position that transcends its historical context implied that the writing of history drew instead on values relative to cultures and individuals. In mathematics, the publication of the paradoxes in set theory earlier in the century unleashed a debate, a "foundations crisis" (*Grundlagenkrise*) over the philosophical foundations of mathematics, which by the late 1920s had already entered philosophical parlance with no sign of resolution.[10] And in 1933, amidst the growing catastrophe of Nazism in neighboring Germany, Hans Hahn, an Austrian mathematician of Jewish descent, diagnosed a "crisis of intuition" (*Krise der Anschauung*) in mathematics as well, as mathematicians produced results that contradicted the hegemony of visual intuition.[11] Just a few years later, by the time Husserl delivered his lectures in Paris and Prague, the implications and potential consequences of this latest crisis in Enlightenment thought—in terms of politics, reason, and the relationship between the two—had become much more severe.

For Husserl, the crisis in the European sciences was no less than a crisis in reasonable society as a whole. At stake was "civilization" based on human values and thoughts, "a rational civilization, that is, one with a latent orientation to reason."[12] The creation of such a rational civilization had been the initial, utopian hope of a universal, mathematical science—the dream of a means to calculate all thought as if it were mathematics in Gottfried Wilhelm Leibniz's *characteristica universalis* and of a unified science of nature and culture in Francis Bacon's *scientia universalis*.[13] Indeed, as Husserl notes, such hopes manifested themselves in the eighteenth century as the Enlightenment sought to reform education, society, and political life. But in 1935, a time far removed from the Enlightenment, Husserl's alarm pointed to a shift in what the sciences meant in Western European society: due to the prosperity that they had produced, the mathematical natural sciences had become the "total world-view of modern man," culminating in an "indifferent turning-away from the questions which are decisive for a genuine humanity."[14] Husserl asserted that, instead of fostering reflection on the value and meaning of human existence, a pressing matter in 1930s Germany, the "fact-minded sciences [made] merely fact-minded people."[15] In 1917, the German sociologist Max Weber resigned himself to the idea that science no longer offered insight into the general conditions of modern life.[16] Two decades later, the ever-worsening political situation meant, for Husserl, that the task of philosophy now lay in locating and correcting the moment at which this totalized, scientific worldview had gone astray.

In Husserl's eyes, this crisis represented not a sudden change in how people understood humanity as ushered in by the rise of authoritarianism in Germany, but rather a change that had taken root centuries before, in the work of Galileo. What was decisively new with Galileo was the idea that the limited application of geometry to astronomy could be extended to the world as the *"mathematization of nature,"* in which *"nature itself* is idealized under the guidance of the new mathematics; nature itself becomes—to express it in a modern way—a mathematical manifold."[17] Galileo's transformation of nature into mathematics thus became the success story of the modern sciences. What worried Husserl, however, was the epistemological and methodological transformation implied by the mathematization of nature, a change driving the crisis of knowledge in the 1930s: "We must note something of the highest importance that occurred even as early as Galileo: the surreptitious substitution of the mathematically substructured world of idealities for the only real world, the one that is actually given through perception, that is ever experienced and experienceable—our everyday life-world. This substitution was promptly passed on to his successors, the physicists of all the succeeding centuries." One feels the urgency in Husserl's tone—the *only* real world; the text continues: "What was lacking, and what is still lacking, is the actual self-evidence through which he who knows and accomplishes can give himself an account, not only of what he does that is new and what he works with, but also the implications of meaning which are closed off through sedimentation or traditionalization, i.e. of the constant presuppositions of his own constructions, concepts, propositions, theories."[18] A number of elements in these two passages resonate with contemporary readers, especially with critical theorists. We register not only a deep ambivalence toward the total mathematization of "our everyday life world" but also how this foundational shift exchanges knowledge as the comprehension of meaning in "the only real world" for inquiry into mathematized nature. Moreover, as this change passes from physicist to physicist we recognize the reification of this unspoken shift, the transformation of historical choices into the way we interpret nature itself. For Husserl, this link between the mathematization of nature and the ever-worsening situation in Germany and across Europe was not explicitly causal. Instead, it provided the conditions to understand and, potentially, return to and correct the point at which we began to foreclose the investigation of our everyday life world and the consequences of that for humanity.

For Horkheimer and Adorno, however, the connection between the mathematization of nature and the crises in Europe in the twentieth century was causal. By the time Husserl died in 1938, Horkheimer and Adorno were already living in the British and American exile from which they wrote texts foundational to the canon of critical theory: "Traditional and Critical Theory" ("Traditionelle und Kritische Theorie," 1937), *Dialectic of Enlightenment* (*Dialektik der Auflärung*, 1947), and *Minima Moralia* (1951). For these two exiled German-Jewish philosophers and social theorists, the mounting catastrophe in their former homeland was not the result of a deviation from the core questions of reason but the *product* of reason, of so-called enlightened society itself.[19] As Galileo symbolized this transformation for Husserl, Francis Bacon personified in *Dialectic of Enlightenment* the duality of reason. He exemplified "the scientific temper that followed him. The happy marriage between human understanding and the nature of things that he had in mind is patriarchal: the understanding, which conquers superstition, is to rule over demystified nature."[20] This linkage was the troublesome promise of enlightenment—taken to mean reason (*Vernunft*) as well as the Enlightenment as a historical period; both forms of enlightenment supplant mythological explanations of the world, but do so in a way that violently subordinates nature in order to control it. For the first generation of critical theorists, the technology of cinema embodied this ambiguous potential of modernity and enlightenment.[21] Mathematics did too, offering the cognitive tools that expanded not only knowledge but also domination from the historical Enlightenment to the present day: "Before and after quantum theory, nature is what can be grasped mathematically; even what cannot be assimilated, insoluble and irrational, is fenced in by mathematical theorems."[22] Like the compass, the cannon, and the printing press, mathematics became an instrument with which reason could formulate, calculate, and, hence, control the world and all that exists in it.

The emergence of critical theory in the works of Horkheimer and Adorno thus shifted how theoreticians of culture and art thought about mathematics. For Horkheimer and Adorno, the proposed equation of mathematics with thought by the logical positivists in the 1920s represented the most recent example of the return of enlightenment to barbarism and violence, exemplified by Odysseus's self-restraint to hear the Sirens' song, Bacon's equation of knowledge with power, and the culture industry's manipulation of the masses. In the period of Horkheimer and Adorno's self-staging of critical theory in the 1930s and 1940s, an intellectual narrative emerged that saw in mathematics not the emancipation, knowledge, and freedom once promised by enlightenment, but rather its relapse into restriction, coercion, and subjugation. This is how mathematics appears in Horkheimer and Adorno's collaborative work: "With the forfeiture of thought, which in its reified form as mathematics, machine, and organization exacts revenge on humans forgetful of it, enlightenment renounced its own realization. By subjugating all particulars to its discipline, it [enlightenment] granted the uncomprehended whole the freedom to fight back as mastery over things against the being and consciousness of humans."[23] By 1935, mathematics pointed Husserl to the aborted realization of the Enlightenment evident in a crisis of reason that materialized for Horkheimer and Adorno in their forced exile. By the end of World War II, the mathematization of thought and nature had become a central factor in answering the question of why "humanity, instead of entering a truly human state" was "sinking into a new kind of barbarism," which the destruction of Europe and the attempted annihilation of the European Jews only confirmed.[24] Indeed, Horkheimer and Adorno's association of mathematics with a regressive vision of thought became an enduring mode of presentation for critical theorists such as Herbert Marcuse and Jürgen Habermas, for whom mathematics symbolized naïve positivism and a mode of social and economic conformity. In the earliest phase of critical theory's development and deployment, the choice facing modern thought seemed clear: either it could expose and resist societal mechanisms of control and domination, an assignment called critical theory, or it could continue to mimic the expedient symbols and operations of mathematics, seemingly indifferent to the fate of humanity.

And yet even the briefest look back into critical theory's intellectual origins, let alone the ideas and letters of the broader German-speaking world in the early twentieth century, challenges the narrative that mathematics must work in opposition to the concerns of humanity. For instance, whereas for Husserl the mathematization of nature vanquished reason from reality, for Siegfried Kracauer the mathematical study of space—geometry—bridged the void between materiality and pure reason. In Kracauer's essays written during the Weimar Republic, the material logic of mathematics informed his readings of mass culture, which sought to advance, rather than oppose, the project of the Enlightenment. For him, geometry enabled a literary approach to cultural critique in which the work of the critic helped confront the contradictions of modernity and, through such confrontation, potentially resolve them. Whereas for Horkheimer and Adorno the mathematization of thought typified the return of enlightenment to barbarism, for Gershom Scholem the philosophy of mathematics dealt with the problem of language at a moment of cultural crisis by omitting representation. This exclusion revealed, at least to Scholem, configurations of language that captured historical and religious experiences whose extremity exceeded language's limits. Following mathematics' lead, restricting representation in poetic language symbolized, as a negative aesthetics, the inexpressibility of the privations of life in exile. And, for Franz Rosenzweig, infinitesimal calculus circumvented the enigma of the infinite, revealing a messianism that brought the messianic moment into the here and now. "Mathematics,"

Rosenzweig writes in *The Star of Redemption* (*Der Stern der Erlösung*, 1921), "is the language of that world before the world."[25] Where empire and war had dissolved the relationships among God, the human, and the world, the austere symbols and mute signs of mathematics offered Rosenzweig a means to reformulate their interconnections. Rosenzweig's messianism and messianic theory of knowledge made human action, belief, and critical thought the motors of achieving emancipation in the real world, restoring to them the same epistemological significance as mathematics.

By tracing this as yet unacknowledged lineage of critical theory, this book explores the underdeveloped possibilities that mathematics held—and still holds—for theories of culture and art. Thanks to contemporary scholars such as Martin Jay, Andrew Feenberg, and Susan Buck-Morss, we know that the intellectual origins of critical theory lie in Sigmund Freud and psychoanalysis, in George Lukács's adaptation of the concept of reification from Karl Marx, and in Walter Benjamin's theorizations of language.[26] I wish to build on these histories of critical theory by returning to the origins of the critical project and recovering a critically productive vision of mathematics in the work of Scholem, Rosenzweig, and Kracauer. This is about more than thinking of the intersections of mathematics, culture, and art in terms of the apparent aesthetic beauty and elegance of a mathematical proof, for example, or of the historical moments at which artists have drawn inspiration from the abstractness of mathematics.[27] In the works of these German-Jewish thinkers, two much more complicated intellectual visions of the relationships among mathematics, culture, and art emerge: one vision—in Horkheimer and Adorno's early vision of critical theory—that sees in mathematics the destructive force of reason and another vision that, about two decades earlier, finds in mathematics methods of navigating the modern crises of the Enlightenment. One of the primary claims of this book is that revisiting these intellectual narratives enables us as critical theorists to rethink how we approach mathematics—not as an antithesis to humanistic inquiry, but instead as a powerful and timely mode of intervening in the worlds of culture and aesthetics.

Defining a Program of Negative Mathematics

For Scholem, Rosenzweig, and Kracauer, the very austerity and muteness of mathematics revealed pathways through apparent philosophical impasse, a chance to realize the Enlightenment's promise of inclusion and emancipation as it seemed to disappear in early-twentieth century Germany. Building on the thought of these three lesser-known German-Jewish intellectuals of the interwar period, I propose an understanding of mathematics that can help move past today's debates that pit the humanities against the sciences. By locating in mathematics a style of reasoning that deals productively with that which cannot be fully represented by language, history, and capital— what I call *negative mathematics*—the work of these three German-Jewish intellectuals illuminates a path forward for critical theory in the field we know today as the digital humanities. Here *negative* mathematics refers neither to the concept of negative numbers nor to the infamous image of Adorno and other members of the Frankfurt School as unremitting naysayers. Instead, it offers a complement to the type of productive negativity that Adorno in particular located in the Hegelian dialectic.[28] We can think of negative mathematics as *negative* in terms of mathematical approaches to issues of absence, lack, privation, division, and discontinuity. One example of such negativity that we will repeatedly encounter in this book is how mathematics develops concepts and symbols to address ideas that, in some accounts, human cognition and language cannot properly grasp or represent in full, such as the concept of the infinite or even the nature of mathematical objects themselves. For Scholem, Rosenzweig, and Kracauer, these mathematical approaches to negativity provided the generative spark for theorizing culture and art anew, where inherited modes of philosophical and theological thought no longer applied to modern life. Negative mathematics thus expands Horkheimer and Adorno's critical project in spite of themselves, introducing avenues for critical thought that treat mathematics as a crucial cultural and aesthetic medium.

In the work of Rosenzweig, Kracauer, and Scholem, negative mathematics provided a progressive yet critical approach to cultural crises as the secularization of the Enlightenment threatened the particularity of religious life and the rationalization of capitalism exchanged aesthetic experience and political action for mass entertainment. It was a shared discourse that saw in mathematics' approach to negativity modes of cultural analysis and intervention. While never a cohesive school or doctrine, we can think of negative mathematics, in the words of Anson Rabinbach, as "an ethos in the Greek sense of a characteristic spirit or attitude (*Haltung*)."[29] As an intellectual ethos, negative mathematics was critical, in the Kantian sense of the term, in that it sought to address and correct the shortcomings and contradictions of reason manifest in language, religion, and mass culture.[30] But negative mathematics was also critical in the sense in which Horkheimer redefined the term in the 1930s: For Scholem, Rosenzweig, and Kracauer, negative mathematics meant "not just the proliferation of knowledge, but rather the emancipation of humans from enslavement."[31] By examining concepts such as language and redemption, negative mathematics allowed these thinkers to refashion them in order to take account of experiences, beliefs, and perspectives otherwise marginalized by mainstream society. Negative mathematics emerged in the brief yet profound window of cultural activity between the World Wars in Germany, at a point when the prospect of realizing an inclusive, self-reflective society—the goal of the Enlightenment—still seemed to exist. The approach faded to the margins of critical theory as mathematics became, in the work of Horkheimer and Adorno, a key accomplice in the return to superstition and violence that was the catastrophe of the twentieth century. In passing over Horkheimer and Adorno's equation of mathematics with barbarism, critical theory continues to forfeit mathematics as a tool not only to understand but also to act in contemporary society.[32] In the age of quantification and big data, negative mathematics thus helps us confront what remains a priority for critical thought: the critique of and intervention in a digital world through critical analysis that succumbs neither to the naiveté of scientific positivism nor the rejectionism of critique.

Mathematical approaches to negativity have a long history in German-Jewish intellectual life and letters that dates back to the Enlightenment itself and sets the stage for the interventions of Scholem, Rosenzweig, and Kracauer in the interwar period. This prehistory begins with the Enlightenment philosopher Moses Mendelssohn for whom mathematics offered a justification for metaphysics. His essay "On Evidence in the Metaphysical

Sciences" ("Abhandlung über die Evidenz in metaphysischen Wissenschaften," 1764) argues that mathematics shares with philosophy its mode of analysis, which makes "obscure and unnoticed" parts of concepts "distinct and recognizable" by unpacking and expounding them through chains of inference.[33] Yet whereas mathematics finds impartiality, in that one easily "grasps" (as in the German *fassen*) its deductions, the truths of philosophy are muddled by the prejudices of the human mind. The essay concludes that "metaphysical truths are capable, to be sure, of the same certainty as mathematics," even if they are not capable "of the same perspicuity [*Faßlichkeit*] as geometric truths."[34] Mathematics thus helped Mendelssohn show that metaphysics rested on stable footing, even if some still refused to accept the validity of its claims. The eloquence of this argument won Mendelssohn the Prussian Academy of the Sciences essay prize in 1763, which helped him gain permission to reside permanently in Berlin—"an unprecedented triumph," writes Alexander Altmann, "for the son of the ghetto who had arrived in Berlin only twenty years earlier."[35] For Mendelssohn, and for a number of German-Jewish intellectuals that followed him, mathematics was a point of entry into debates about metaphysics and reason that signified not only a powerful philosophical tool but also a means of inclusion, allowing those of Jewish heritage to participate in the society and culture of the Enlightenment.

For Salomon Maimon, another Jewish philosopher of the German Enlightenment, the latest developments in mathematics intervened in a central debate of the times: the nature of pure reason. As a commentary on Kant's critical philosophy, Maimon's *Essay on Transcendental Philosophy (Versuch über die Transcendentalphilosophie*, 1790) agreed with Kant that the mind plays an active role in constituting the contents of thought, but Maimon claimed that pure reason must originate in thought itself and not draw on the world of experience, as Kant had suggested.[36] According to Maimon, we can think of the pure generation of thought as following not from experience but rather from the intellectual tools employed in infinitesimal calculus that Leibniz and Newton had developed in the previous century to calculate motion in the new mechanics.[37] Their calculi hinged on the idea of infinitely small increments that Leibniz had called differentials; these infinitesimal quantities allowed Leibniz to calculate the rate of change of a curve. For Maimon, the differential provided the origin of pure cognition as a medium between experience and thought. "Sensibility," he writes, "provides the differentials to a determined consciousness; out of them, the imagination produces a finite (determined) object of intuition; out of the relations of these different differentials, which are its objects, the understanding produces the relation of the sensible objects arising from them."[38] Reason appeals not to experience, but rather to how the differentials present experience to the mind as a set of relations, out of which thought can construct pure knowledge. For Maimon, mathematics bridged the seeming impasse between experience and transcendental philosophy but signified more than just an interjection into an ongoing philosophical debate. Alongside the natural sciences, mathematics had played a key role in Maimon's decision to move from a life governed by Jewish orthodoxy in the provinces of Polish Lithuania to an "emancipated" life in cosmopolitan Berlin.[39] Indeed, mathematics allowed him not only to sustain himself in Berlin as a tutor but also to participate in the city's enlightened circles through his *Essay on Transcendental Philosophy*.

Almost a century later, mathematics again provided the keys to pure thought for the German-Jewish philosopher Hermann Cohen. The embodiment of the post-Enlightenment spirit of a Jewish synthesis with German culture and the hope for a truly egalitarian Germany, Cohen was the first Jew to hold a full professorship in Germany.[40] Philosophy, he advocated, must be saved from Hegelian speculation via a return to the Kantian tradition of idealism, taking mathematics as the basis for a scientifically grounded metaphysics. As it had for Maimon, infinitesimal calculus offered a method of generating the objects of pure thought without recourse to intuition and experience.[41] In *The Principle of the Infinitesimal Method and its History* (*Das Princip der Infinitesimal-Methode und seine Geschichte*, 1883), Cohen asserts that pure thought creates the continuous fabric of metaphysical reality (*Realität*) in the same fashion that, in mathematics, infinitesimal tangent lines can be thought of as producing a curve.[42] The mathematical genesis of the contents of cognition became, in the logic of Cohen's *System of Philosophy*, the foundation of the Neo-Kantianism that shaped the German philosophical academy around 1900: "The analysis of the infinitesimal is the legitimate instrument of the mathematical natural sciences. . . . This mathematical generation [*Erzeugung*] of movement and, thereby, nature is the triumph of pure thought."[43] To Cohen, mathematics—and in particular the watershed mathematical developments of the Enlightenment that made the Newtonian cosmos knowable through calculation and prediction—provided the conditions of possibility for pure thought.

In Cohen's work with mathematics, I recognize something new that would be pivotal for Scholem, Rosenzweig, and Kracauer: a link among mathematics, negativity, and theories of culture and religion. For Cohen, mathematics represented the possibility of pure knowledge that underpinned his concept of a religion of reason, as derived in his posthumously published work *Religion of Reason Out of the Sources of Judaism* (*Religion der Vernunft aus den Quellen des Judentums*, 1919). Here Cohen drew on his earlier work, *The Logic of Pure Knowledge* (*Die Logik der reinen Erkenntnis*, 1902), in which infinitesimal calculus rendered legible and scientifically operative the pure genesis of thought, "the judgment of origin." Accordingly, thought originates not in the negation of something ("A" is not "nothing") but rather in the determination of the positive, infinite possibility for what something ("A") is not ("nothing"), exemplified by the concept of the infinitely small in mathematics.[44] In *Religion of Reason*, this mathematical origin of pure thought provided the terms for the pure cognition of God's attributes. Drawing on the medieval Jewish philosopher Maimonides, Cohen's final work argues that we have positive knowledge about God through the judgment: "God is not inert [*träge*]."[45] In the same fashion that the finite line originates in the infinitesimal point in mathematics, we can think of God as the infinite totality of activity, all that which is not inert and inactive ("träge"). Along with Mendelssohn's and Maimon's arguments, Cohen's usage of mathematics here is remarkable. Not just metaphysics but also cultural discourse on religion, if they are to draw on reason, require a method that is both logically certain and self-evident, apodictic and exemplary, and only mathematics fulfills the duality of this task.

Here, I do not devote separate chapters to Mendelssohn, Maimon, or Cohen. Instead, I take their mergers of mathematics with metaphysics and of mathematics with religious and cultural thought, as well as the intellectual possibilities that these mergers opened up for German Jews, as points of departure for negative mathematics in the interwar period.

Mathematics, Metaphor, and the Experience of Modernity

This book argues that the contributions that Scholem, Rosenzweig, and Kracauer made to the project of critical theory become legible in the thinkers' deployment of specific sets of metaphors that they drew from mathematics and mathematical approaches to negativity. These sets of metaphors depended on and reflected the diverse branches of mathematics from which they were drawn—a diversity often obscured by the singular and seemingly monolithic abbreviation, *math*. For Scholem, the philosophy of mathematics signified purity, privation, and structures of language lacking representation. For Rosenzweig, infinitesimal calculus implied motion over rest (the absence of motion) and a form of subjectivity that dynamically grasped the otherwise unknown elements of the physical world. And, for Kracauer, geometry pointed to the concept of space as a bridge across the void separating experience and cognition. Although these metaphors may not all directly embody negativity, their common link to negativity lies in the fact that they were derived from mathematical strategies for dealing with issues of lack, absence, and privation. Signifying mathematical approaches to negativity made these metaphors applicable when issues of negativity became manifest in the cultural and aesthetic sphere. For Scholem, Rosenzweig, and Kracauer, the metaphors of negative mathematics uncovered the deeper dimensions and illuminated the dark corners of language, redemption and eternity, and the tenuous link between materiality and cognition that, in their work, translated into strategies to confront the intellectual impasses presented by the early twentieth century.

These metaphors represent the critical potential of negative mathematics, but their status as metaphor has also conditioned the exclusion of mathematics from critical and scholarly discourse. As a more delicate interaction between cultural and mathematical thought, the function of mathematical metaphors has been overshadowed, in part, by the polemic equation of mathematics with a restrictive and limited mode of thought by Horkheimer and Adorno's inception of critical theory and by their subsumption of mathematics into a narrative of enlightenment's dialectical return to myth and barbarism. But we as scholars have also missed the significance of these mathematical metaphors, because we have viewed them in the pejorative sense as just that—as metaphors, analogies, the remnants of inauthentic speech.[46] To grasp the critical potential of negative mathematics, we must think of mathematics less as a simple and limited analogy for a formal methodology and more as a conscious and consequential rhetorical strategy. For Scholem, Rosenzweig, and Kracauer, negative mathematics was not insight provided by some mathematical theorem nor did it function in terms of the figurative power of a lone metaphor. Instead, negative mathematics operated in terms of the conceptual implications that interconnected and governed cohesive sets of metaphors, "metaphorics" to adopt the German term (*Metaphorik*), drawn from different branches of mathematics. As a metaphorics, the modes of mathematical thinking unique to the philosophy of mathematics, infinitesimal calculus, and geometry corresponded to the distinct influence that each branch of mathematics had on Scholem, Rosenzweig, and Kracauer, resulting in, respectively, an aesthetics of privation, a dynamic messianism, and a materialist form of cultural criticism. As systematic sets of metaphors and not as simply analogies for formalized thought, these metaphorics served as the medium in which ideas could transfer between mathematics and theories of culture and aesthetics.

Paying closer attention to these metaphorics, then, enables us to recover the specific contribution that negative mathematics made for these German-Jewish intellectuals. In taking this approach, I draw on Hans Blumenberg's study of metaphor, *Paradigms for a Metaphorology* (*Paradigmen zu einer Metaphorologie*, 1960), which views metaphors in philosophical discourse as more than just the "leftover elements" of the process of creating philosophy's clear and distinct concepts. For Blumenberg, philosophical metaphors and, in particular, our study of them, "brings to light the metakinetics of the historical horizons of meaning and ways of seeing within which concepts undergo their modification."[47] Blumenberg discusses, for example, how a metaphorical shift in the concept of truth precipitated the rise of the modern experimental sciences: the medieval notion of the "mighty" truth required truth to overpower the passive knowing subject, whereas the "hidden" truth of the modern period necessitated the labor of the active intellect to discover, experiment, and know it.[48] Likewise, in my readings of Scholem, Rosenzweig, and Kracauer, it was through the metaphorics that arose around mathematical approaches to absence, lack, and discontinuity that mathematics impacted and shaped cultural and aesthetic discourse. This book charts the systematic construction and theoretical consequences of metaphors of purity and privation in the philosophy of mathematics, metaphors of motion and subjectivity in infinitesimal calculus, and metaphors of space in geometry. Tracing the implications of mathematical metaphors in the work of these German-Jewish thinkers reveals the moments where mathematics' approach to negativity expanded the horizons of cultural and aesthetic thought to include minoritarian perspectives, such as ideas that evade representation and the histories of marginalized groups.

For Scholem, Rosenzweig, and Kracauer, negative mathematics intervened at a particularly precarious moment in Jewish intellectual existence in Germany during the early twentieth century. Indeed, the emergence of the metaphorics of negative mathematics in their thought coincided with the end of World War I and the collapse of Imperial Germany, experienced as a world-historical destabilization of philosophical and political authority accompanied by the crises and freedoms that such destabilization afforded.[49] These crises and the sense in the early twentieth century that Jews, despite their legal emancipation in 1812, had never become full members of German society called into question the theological and philosophical modes of social and cultural engagement inherited from the generations of Mendelssohn and Cohen.[50] Amid the growing unease of cultural crisis, negative mathematics showed Scholem, Rosenzweig, and Kracauer paths through these modern crises by offering ways of reconfiguring language, history, messianism, and cultural criticism that worked to realize the emancipatory promise of the Enlightenment. The story of negative mathematics and critical theory is, in other words, a German-Jewish story, not only because the majority of the protagonists were born in Germany of Jewish descent and worked primarily in the German language. It is also a German-Jewish story because the mathematical metaphors developed by these authors addressed concerns of reason, inclusion, and, ultimately, exile and extermination tied to the historical experiences of Jews living in Germany. At stake in investigating negative mathematics, the origins of critical theory, and German-Jewish intellectual life are ways of not only pulling apart inherited philosophical, theological, and cultural categories but also redefining more inclusive visions of them, which Horkheimer and Adorno's opposition of mathematics and critical theory has tended to eschew.

One of the primary critical categories further illuminated by the meta-phorics of negative mathematics is the persistent theological dimension of critique, which interlinks the critical project with an insistence on redemp-tion—or, at least, an insistence on the need for redemption. For Scholem, Rosenzweig, and Kracauer, critique was bound up with a refusal that this world, as broken as it appears and is, is all that there is. Since the first histo-ries of the critical project, scholars have emphasized and built upon the "weak messianic" element in critical theory, operative most notably in the work of Adorno and Benjamin.[51] Even after Auschwitz, critique held hope for the possibility of radical change in the historical process that, especially for the first generation of critical theorists, was located in aesthetics.[52] The meta-phors of negative mathematics, however, reveal the presence of mathe-matics in this theological register and suggest that mathematics' differing approaches to negativity can help excavate the deeper layers of critique's theological impulse. Mathematical logic, for instance, allowed Scholem to rescue marginalized, precarious, and often unspeakable experiences and tra-ditions, such as Judaism, from erasure and oblivion. While scholars have started to pay closer notice of theological concerns of Adorno's materialism, infinitesimal calculus already served Rosenzweig as a way of attending to and refusing to give up on even the infinitesimal and seemingly insignificant aspects of life, history, and the world.[53] Finally, the synthesis of materiality and logic in geometry pinpointed perhaps the most tireless theological as-pect of the critical project: the idea that the practice of critique is itself a fundamentally redemptive enterprise, centered on the possibility of rea-son's intervention into the material conditions of life. As the following chap-ters explore the development and deployment of negative mathematics in Scholem, Rosenzweig, and Kracauer's work, readers will recognize these mathematical imprints on the theological dimension of critical thought—not as attempts to expand knowledge for knowledge's sake, but rather as a means of articulating the emancipatory potential of critique.

Ultimately, the fact that mathematics enabled these German Jews to the-orize critical yet also inclusive visions of culture and art points to the rele-vance of negative mathematics for critical theory in the digital age. For Scholem, Rosenzweig, and Kracauer, negative mathematics was able to re-integrate into theory Jewish perspectives on history, redemption, and cultural critique because it was—and still is—concerned with and essential to a core set of cultural questions and anxieties over the fate of language, the viability of critique, the nature of reason, and the path toward societal emancipation and recognition. Negative mathematics thus helps us in the present conceive of theory that takes advantage of what mathematics (and technologies based on mathematics and quantification) offer analytically, while remaining committed to the critical project. The fact that negative mathematics helped these German-Jewish intellectuals to push past cultural impasses of the 1920s suggests that the potential of digital technologies also lies in allowing underrepresented and marginalized communities, those living in exile, or peoples with more oblique relationships to power to break up and redefine social and cultural categories. Negative mathematics implies that the critical potential of digital humanities lies, as practitioners such as Lauren Klein have shown, in exposing and giving voice to the silences, discontinuities, and modes of exclusion that remain, in part because of digital technologies, in contemporary society.[54] To be sure: recovering negative mathematics is not a return to a once happy marriage between mathematics and cultural and aesthetic theory. Reinstating the productive tensions between mathematics and critical theory—as often competing but not necessarily opposed ways of approaching the cultural problems of the present—is the goal of this book.

Overview of the Book

The following four chapters examine the emergence of the intellectual tensions between mathematics and critical theory and explore negative mathematics as an alternative paradigm for thinking about mathematics in cultural and aesthetic thought. Chapter 1 investigates the construction of a seeming opposition between mathematics and critical theory as first framed by Horkheimer and Adorno in conversation with Benjamin and Lukács. This opposition emerged out of an acrimonious philosophical confrontation with members of the Vienna Circle and their vision of a scientific philosophy, logical positivism. Mathematics served as the grounds on which this disagreement played out, as the logical positivists' reliance on mathematics linked them for the critical theorists to a form of political quietism and the acceptance of authoritarian government. At stake were, at least for Horkheimer and Adorno, not only the concepts of subjectivity, experience, and language but also the course and political imperative of modern philosophy. The critical theorists' side of the debate cast mathematics as instrumental reason, reification, and a restricted form of thought that perpetuated the status quo amidst the increasingly troublesome political situation of the late 1930s. What is striking—and what has been pervasive about the image of mathematics set into motion by Horkheimer and Adorno—is how it transformed a historical intellectual conflict into a history of thinking that associated mathematics with the breakdown of reason that brought about Hitler and Auschwitz. Even after the war, the image of mathematics established in this brief but decisive phase of critical theory's inception persisted as a symbol of positivistic thinking and a tool of societal control.

Subsequent chapters return to the origins of the critical project before Horkheimer and Adorno's confrontation with logical positivism to reconstruct the creative possibilities for critical thinking revealed by negative mathematics, the intellectual project shared by Scholem, Rosenzweig, and Kracauer. Chapter 2 investigates the first formulation of this project in the work of Scholem. For Scholem, the debates surrounding the philosophical foundations of mathematics revealed the aesthetic and historical potential of privation: mathematics, in particular mathematical logic, produces novel results by abandoning the conventional representational and meaning-making functions of language. The philosophy of mathematics provided Scholem with metaphors of structure lacking the representational functions of language. These metaphors enabled him to counter not only the previously mentioned crisis and skepticism that surrounded language but also his growing sense of the unviability of Jewish emancipation and equality in Germany. In his work on mathematical logic, these metaphors of privation opened up for Scholem unlikely avenues of aesthetic and linguistic expression, showing how language as silence can serve as a symbol of its own limitations. How mathematics deals with the shortcomings of language showed Scholem the expressive potential of the poetic genre of lament and informed his translations of biblical lamentations, laying the groundwork for a philosophy of history that underpinned his *Major Trends in Jewish Mysticism* (1941). Like mathematics, lament and history mobilized for Scholem the idea of privation—the experience of deprivation—to symbolize that which remains unsayable in language and untransmissible in history. Scholem's negative mathematics thus bears the possibility for theories of culture and history that could account for the experience of exile and diaspora, finding historical transmission and continuity in moments of silence, rupture, and catastrophe.

Rosenzweig's mathematics-inflected intellectual program materialized around the same time as Scholem's poetics of negation, but the former focused on embedding messianism into the everyday work of thought. In chapter 3, I revisit the debates surrounding Leibniz and Newton's calculi in the philosophy of Cohen and trace how the concept of the infinitesimal quantity (the differential) signaled to Rosenzweig not only the source of pure cognition but also metaphors of motion, rest, and the primacy of the former over the latter. Benjamin Pollock has shown that Rosenzweig drew on German Idealism and Cohen's use of the differential to rebuild thought in response to crises unleashed in 1914.[55] But practical, contemporary pedagogical debates and popular intellectual histories emphasizing the significance of infinitesimal calculus and the concept of the differential influenced Rosenzweig and his *Star of Redemption* as well. Infinitesimal calculus revealed the ways in which thought could account for the actuality of motion as it is experienced in the world by synthesizing finitude in reference to infinitude in a single concept, the differential. *The Star of Redemption* hinges on this approach as it transforms the finite, thinking subject into the agent of revelation and an active participant in the creation of the eternal Kingdom of God on earth. This is the messianic role that the individual assumes in Rosenzweig's theory of knowledge, the "New Thinking," which makes room, alongside the truths proved by mathematics, for the truths verified by the beliefs of individuals in the course of history—including historically marginalized groups, such as the Jews in Germany. It is also a messianism and an epistemology that informs critical theory, where the cultural critic works in the dim light of messianic reconciliation.

With aims equally as grand as Rosenzweig's messianism, the style of cultural critique cultivated by the journalist and philosopher Siegfried Kracauer employed negative mathematics as a means of working toward a society based on reason through the aesthetic composition of his analyses. Chapter 4 deals with the metaphorics of space and the method of projection that Kracauer found in geometry. In his early theoretical texts and feuilletons for the *Frankfurter Zeitung*, mathematics fulfilled an impossible yet pressing assignment: in a world vanquished of authority, geometry and the metaphorics of space bridged the divide between the raw contingency of materiality and the necessity of a priori laws. Drawn from his training as an architect, geometry also provided Kracauer with the analytic method of projection, which read in the material products of mass culture the metaphysical trajectory of history and the modern crisis point represented by capitalism. These rationalized products of capitalism (detective novels, dance revues, etc.) embodied this troubling stagnation of Enlightenment reason into mere rationality, and yet, according to Kracauer, if we confront rather than ignore the petrification of reason we may still realize the true progressive force of the Enlightenment. As a "natural geometry," the metaphorics of space and method of projection became a literary strategy for Kracauer, an aesthetic styling of his texts, which, in my account, served as a programmatic attempt to stage publically a confrontation of Enlightenment reason with capitalist rationality. For Kracauer, this was the unique task of the marginal figure of the societal observer, the Jew, the cultural critic, who thus played a salient role in correcting the historical trajectory of reason. Even as mathematics came under scrutiny with the rise of Fascism, Kracauer's claim that the aesthetics of theory work toward a societal confrontation with rationalization suggests that the critical project ought to take seriously the material and performative dimension of criticism as a mode of cultural intervention in a still hyper-rationalized and, now, digitized present.

Around the end of the 1920s, mathematics appeared to lose its critical appeal, transforming, through the work of Horkheimer and Adorno, into an instrument of oppression and totalitarianism. And yet mathematics has by no means disappeared from our philosophical and cultural horizons. The conclusion considers the persistence of the intellectual positions and antagonisms of the past century in contemporary debates over the place of mathematics, quantity, and computation in the humanities. Based on mathematical processes, computational approaches to the humanities—known as the digital humanities—offer broader access to cultural and aesthetic products and new insights into their composition, circulation, and interrelation. However, those skeptical of the incursions of quantitative and computational methods into humanistic inquiry recognize that proponents of such mathematically inflected methods all too often reiterate a scientific optimism and rejection of the humanistic tradition akin to that of the logical positivists. Like Horkheimer and Adorno, contemporary critics of the digital humanities claim that digital humanists focus too narrowly on technology and code at the cost of politics and language, history and critique. Negative mathematics offers a third way for the digital humanities to maneuver between an uncritical positivism and the rejectionist impulse of critical theory. Drawing on the ideas offered by Scholem, Rosenzweig, and Kracauer's project of negative mathematics, the conclusion maps out this third way, which sees in mathematical and computational approaches to negativity ways to capture and express otherwise marginalized experiences, histories, and cultures. Understanding how negative mathematics once helped shape critical theories of culture and art opens avenues for theorizing a more reasonable and inclusive society, avenues that enable the humanities to draw on the analytic benefits of mathematics while, at the same time, reconfiguring the limits of representing minoritarian ideas and peoples in the digital age.

The Trouble with Logical Positivism: Max Horkheimer, Theodor W. Adorno, and the Origins of Critical Theory

This chapter explores the reasons why the potential contributions of mathematics to the critical project have gone overlooked by reexamining the dispute between the Frankfurt School and the Vienna Circle during the 1930s. The debate—fought primarily between Horkheimer and Adorno and members of the Vienna Circle such as Otto Neurath (1882–1945) and Rudolf Carnap (1891–1970)—was a fight over the direction of modern thought in which Horkheimer and Adorno saw the fate of Europe hanging in the balance.[1] Against the backdrop of the rise of Fascism and World War II, the dispute transformed mathematics from a point in a philosophical debate into a key factor in the dialectics of enlightenment. Horkheimer and Adorno's criticism of the Vienna Circle's program of logical positivism also served as an initial self-conception and defense of a vision of critical theory opposed to the epistemological primacy that the logical positivists afforded mathematics. For Horkheimer and Adorno, emphasizing mathematics at the cost of language, subjectivity, and a deeper understanding of metaphysics threatened to eliminate concepts from philosophy that held its last hope for intervening in an ever-worsening political climate. While fathomable within its historical context, the seeming opposition between critical theory and mathematics that emerged out of this debate was neither a necessary nor a foregone conclusion. Instead, this opposition between mathematics and critical theory was the byproduct of a conscious and concerted attempt on the part of Horkheimer and Adorno to establish and preserve a philosophical legacy, to draw the boundaries between intellectual friends and enemies, not to mention to compete in the tight market for university resources and positions in exile.[2] The result of the phase of critical theory accompanying Horkheimer and Adorno's clash with logical positivism was that the critical contributions made by Scholem, Rosenzweig, and Kracauer through mathematics have gone largely unnoticed.

In essence, the image of mathematics in Horkheimer and Adorno's vision of critical theory that emerged out of this debate held that mathematics helped drive enlightenment's return to barbarism that ended with Hitler and Auschwitz. "Mathematical procedure became, so to speak, a ritual of thought," Horkheimer and Adorno write in their main collaborative work, *Dialectic of Enlightenment* (*Dialektik der Aufkärung*, 1947). "Despite its axiomatic self-limitation, it installed itself as necessary and objective: it turns thinking into a thing—a tool, in its words."[3] In the wake of the dispute with logical positivism, mathematics became for Horkheimer and Adorno one of the "tools" of instrumental reason, personified by the figure of Francis Bacon, which strove to dominate nature (and other humans) in its obsessive quest for self-preservation. In contrast to the method of sound reasoning that the logical positivists found in mathematics, it represented for Horkheimer and Adorno the ritual repetition of logical operations that turned thought into an object, a mythic totem. While scholars of critical theory have documented Horkheimer and Adorno's conflict with logical positivism, I turn here to the period in their thought that subsumed mathematics in their critique of logical positivism and the concept of enlightenment.[4] This period of Horkheimer and Adorno's writing transformed mathematics into a symbol of exclusion and oppression, cutting it out of the investigations of language, experience, and subjectivity that they then called critical theory.

In order to get a sense of the stakes and implications of Horkheimer and Adorno's hostility toward mathematics, it is necessary to take a closer look at the height of the confrontation between the Frankfurt School and the Vienna Circle. In 1937, five years before Horkheimer and Adorno began composing their critical stance on mathematics in *Dialectic of Enlightenment*, Adorno returned to the continent that the National Socialists had once forced him to leave. Along with Walter Benjamin, who was living as an exile in Paris, Adorno was representing the Institute for Social Research on Horkheimer's behest, planning to attend the Third International Congress for the Unity of Science, organized by members of the Vienna Circle including Neurath and Carnap.[5] What Neurath and Carnap said at the Congress interested Horkheimer. He had been working with Neurath to establish collaborative relationships between the institute and the Vienna Circle, but had published a polemical essay attacking logical positivism earlier in 1937, "The Latest Attack on Metaphysics" ("Der neuste Angriff auf die Metaphysik").[6] Instead of discussing the groups' common interest in empirical sociology, the essay sought to expose how logical positivism's "scientific philosophy" represented a "pathetic rearguard action undertaken by the formalistic epistemology of liberalism, which in this area, as in others, turns into open advocacy of Fascism."[7] For the logical positivists, the philosophical turn to mathematical formality worked toward "cleanliness," "clarity," and "rigor" against the resurgence of metaphysics and the fantasies of Nazi politics, striving for "a neutral system of formulae, for a symbolism freed from the slag of historical languages."[8] For Horkheimer and for Adorno and Benjamin in Paris, however, the middle of the 1930s was not the time for neutrality. Instead, such a radical turn to mathematics in philosophy threatened, they believed, the critical capacity of thought as the practice of reason. The stakes of this debate were deeply political: if philosophy became the formulae of mathematics, then, the critical theorists believed, thought would be impotent to resist and, therefore, complicit in the authoritarianism engulfing Europe.

Aside from the turn to mathematics, it may seem at first as if the Frankfurt School and the Vienna Circle had much in common. Both were comprised of leftist intellectuals who were forced to leave Germany and Austria because of their political views, because of their Jewish heritage, or because their philosophy was construed as "Jewish."[9] Members of both schools—notably Horkheimer, Adorno, and Carnap—also shared an intellectual lineage, having studied under the Neo-Kantian tradition of philosophy.[10] For Horkheimer and Adorno, who remained committed to Hegel, Marx, and the dialectic as a philosophical method, the trouble with logical positivism lay in their self-described "anti-metaphysical" and even antiphilosophical combination of empiricism and mathematics.[11] As described in the circle's 1929 manifesto, *The Scientific Conception of the World* (*Wissenschaftliche Weltauffasung*), logical positivism "knows only empirical statements about things of all kinds, and analytic statements of logic and mathematics."[12] For the Vienna Circle, true "science" and "knowledge" (*Wissenschaft*) should consist exclusively of basic, empirically verifiable statements and the symbols and operations of mathematics; other forms of philosophy, including the critical dialectics of the Frankfurt School, represented "not theory or communication of knowledge, but poetry or myth [*Dichtung oder Mythus*]."[13] Horkheimer and Adorno's rejection and criticism of this standpoint served as the basis of the controversy between the Frankfurt School and the Vienna Circle and, by extension, their antagonistic view of mathematics. Indeed, as Horkheimer argues in "The Latest Attack on Metaphysics," excluding language and art from philosophy in favor of mathematics not only turns thought into a tool, an instrument of industry and oppression but also signals a return to "neoromantic metaphysics."[14] Amidst the rise of Fascism, the experience of exile, and Benjamin's forced suicide in 1941, the idea that philosophy would become mathematics at the cost of politics, language, and aesthetics started to frame the fate of modern thought in stark terms. It was a decision between mathematics, which could only perpetuate the status quo of totalitarianism, and dialectics, which as a critical theory worked toward a reasonable society through critique.

In Horkheimer and Adorno's confrontation with logical positivism, three themes emerged that would be formative for not only the relationship between mathematics and critical theory but also for the history and practice of critical theory after World War II: the effects of thought reduced to mathematics, mathematics as reification and instrument of reason, and its role in the collapse of enlightenment. Following Horkheimer and Adorno, practitioners of critical theory tended to understand mathematics in the terms laid out by Weber's thesis of modernity as the disenchantment of the world. Not just philosophy done as mathematics, but mathematics itself underpinned the categories of economic utility, scientific advancement, rationalization, and, ultimately, the domination of nature as a prime example of instrumental reason.[15] So interpreted, mathematics functioned in a societal context as another form of reification, turning relationships between humans into relationships between things—abstract quantities, variables, and equations.[16] This pejorative view of mathematics may be familiar to contemporary critical theorists, but it is only half the story of mathematics in critical theory. Horkheimer and Adorno's encounter with logical positivism also helped them refine some of their signature theoretical concerns—in contradistinction to mathematics: the epistemological significance of language and philosophical style, the contribution of subjectivity to the concept of experience, and the mediated nature of thinking itself. And, as with their totalizing narrative of enlightenment, Horkheimer and Adorno's criticism of mathematics helped shape the postwar landscape of critical theory, overshadowing the contributions that mathematical approaches to negativity, negative mathematics, could make to the critical project.

Mathematical Calculus: The Liquidation of Language and Philosophy

One of the deepest rifts between Horkheimer and Adorno and the logical positivists lay in their differing view of the relationship among knowledge, language, and philosophy. In terms of scientific knowledge, language functioned best, according to the logical positivists' "new logic," as "a neutral system of formulae," an instrument of reasoning and communication based on mathematical logic (*Logistik*) developed by mathematicians and logicians such as Gottlob Frege and Bertrand Russell.[17] Mathematical logic sought to circumvent the ambiguities of "historical languages" by expressing, manipulating, and interpreting logical statements not in the words of German but in the symbols and according to the rules of mathematics. For the first generation of critical theorists, substituting mathematics for language ran counter to the programmatic view on language that Benjamin had laid out in his work on language and translation just over two decades before he and Adorno attended the Congress of the Unity of Science in 1937. For Benjamin, the philosophical and even theological significance of language, the "magic of language," was language's infinitude, which exceeded just its communicative function.[18] As the element that replaced questions of language, materiality, and mediation, mathematics came to signify an incomplete mode of philosophy in Horkheimer and Adorno's confrontation with logical positivism. While mathematics helped define the position of language in critical theory, its seeming opposition to language foreclosed the idea that mathematics could function as a medium of philosophy, at least for the first generation of critical theorists.

To understand this aspect of Horkheimer and Adorno's emergent criticism of mathematics, it helps to review the logical positivists' relationship to language. Recall that one aim of the logical positivists' intellectual program was to limit knowledge to empirical statements and the analytic statements of mathematics expressed through mathematical logic. This restriction intervened in a post-Kantian debate over mathematics and logic. In brief, in the *Critique of Pure Reason*, Kant argued that mathematical judgments are synthetic a priori: they are *a priori* in that we need not make recourse to experience to make them and *synthetic* in that mathematical judgments include new information not already implied by the concepts under analysis. For instance, in the sum $7 + 5 = 12$, I avail myself of an idea (namely, addition) not already contained in the concepts of 7 and 5, relying on intuition to count first 7 and then 5 to get 12.[19] In contrast, logical positivism rejected the existence of "synthetic judgments a priori," emphasizing instead the "conception of mathematics as tautological in character."[20] "Statements in logic and mathematics are tautologies, analytic propositions," Carnap writes, "valid on account of their form alone. They have no content [*Aussagegehalt*], that is to say, they assert nothing as to the occurrence or nonoccurrence of some state of affairs."[21] The logical positivists went so far as to exclude programmatically synthetic judgments a priori from science altogether, discrediting them as metaphysics, poetry, and theology. One can perhaps best capture this dynamic in reference to Ludwig Wittgenstein, whom the Vienna Circle saw as an intellectual ally and forerunner. In 1921,

Wittgenstein famously claimed: "what we cannot speak about we must pass over in silence [*darüber muss man schweigen*]."[22] For the logical positivists, what we could know was what we could express in the seemingly neutral symbols and analytic statements of mathematics, while the rest, the apparent nonsense of metaphysics, was best left unsaid.

For Horkheimer and Adorno, reducing thought to mathematics represented both the philosophical shortcoming and intellectual danger of logical positivism. Responding to Horkheimer's proposed criticism of logical positivism, which became "The Latest Attack on Metaphysics," Adorno claims that this "'new' logic is in truth 'destructive' in precisely that fatal sense of the word, that it simply hands over all content to irrationality and contingency and appeases its need for security with tautologies. I accept your critique word for word. An analysis must proceed, in my opinion, by attacking mathematical logic [*Logistik*] at the point where it still tries to communicate with contents—namely as a theory of language."[23] This passage makes two consequential points for Horkheimer and Adorno's position on mathematics. The first of these is a point that I will return to throughout this chapter: reducing thought to "mathematical logic" sacrifices the "content" of philosophy, its ability to make ethical, aesthetic, and meaningful statements about the world. The problem with this sacrifice is not only its abandonment of the complexities of language and society but also its philosophical impossibility, in as much as mathematical logic, even its most linguistically austere forms, still has recourse to conceptual "contents." Here Adorno's letter passes over how exactly mathematical logic deals with content, which, for Scholem, was the source of mathematics' critical potential (see chapter 2). The second point made by this passage is the expansion of Horkheimer and Adorno's critique to include not only what Horkheimer refers to as the "scientific philosophy" (*wissenschaftliche Philosophie*) advocated by the Vienna Circle but also mathematical logic, which was only a part of the logical positivists' program.[24] Indeed, the passage describes the exchange of "content" for "mathematical logic" and mathematical "tautologies" in dire terms as a "destructive" and "fatal" element of logical positivism. By making this shift, this passage subsumed mathematics and mathematical logic into Horkheimer and Adorno's criticism of logical positivism as a whole.

The critical stance against mathematics that Adorno added to Hork-heimer's critique of logical positivism built on and developed Benjamin's early writings on the relationship between mathematics and language. Benjamin's interest in mathematics paralleled his early friendship with Scholem, then a student of mathematics and mathematical logic.[25] As Peter Fenves shows, mathematics influenced Benjamin's notion that language is infinite, but it also revealed the unique properties of language as systems of representation.[26] Consider briefly Benjamin's discussion of Russell's paradox.[27] What is significant in mathematics is how its statements take the form, according to Benjamin, of a "judgment of designation": one thing "describes" another (as in the German term *bezeichnen*). The mathematical statement "*a* designates the side *BC* of a triangle" posits the symbol *a* as the side of the triangle, *BC*. The subject *a* is only meaningful in relation to *BC* and, thus, cannot enter into "logical relationships" with other objects (the side of square *XY*) without losing its specific meaning (*BC*).[28] In the context of how mathematics represents its objects of study, its judgments constitute "improper meaning" (*uneigentliche Bedeutung*) according to Benjamin, because *a* is meaningless outside of the context in which we decided it would designate *BC*. In contrast, other forms of judgment such as in language retain a likeness to the object that they signify, reminiscent of the divine language of names, in which words and objects correspond, that Benjamin proposed in his 1916 essay "On Language as Such and on the Language of Man" ("Über Sprache überhaupt und über die Sprache des Menschen")."[29] For Benjamin, such judgments have "proper meaning" (*eigentliche Bedeutung*). In language, words represent their object (a tree, for instance), retaining the object's meaning (such as connotations of steadfastness and grandeur) when brought into logical relationships with other judgments.[30] This distinction was important for Benjamin, because signification through "proper meaning" contains representation and, thus, allows for interpretation and criticism. Accordingly, one can explicate and relate judgments in language, while mathematical judgments remain meaningful only within their predefined context.

For Benjamin, these two types of judgments delineated the different spheres of knowledge afforded by mathematics and language. In 1917, Benjamin's "coming philosophy" called for a philosophy that attended specifically to the latter: knowledge contained in language and experience, which the overemphasis of mathematics in the Kantian system had pushed to the margins of philosophy.[31] A decade later, Benjamin set this corrective as the task for *The Origin of German Tragic Drama* (*Ursprung des deutschen Trauerspiels*, 1928), as its "epistemological-critical foreword" begins:

> It is unique to philosophical writing that it must, with every turn, stand anew before the question of representation. Indeed, philosophy will be doctrine [*Lehre*] in its finished form, but it does not lie in the power of mere thought to lend philosophy such closure. Philosophical doctrine rests on historical codification. It cannot therefore be conjured *more geometrico*. The more clearly as mathematics proves that the total elimination of the problem of representation—which every strict didactic system presents itself as—is the sign of true knowledge, the more conclusively does it present its renunciation of the area of truth meant by the languages.[32]

This passage intensifies the association of mathematics with arbitrary "judgments of designation," now finding in mathematics the exemplar of "the total elimination of the problem of representation." At the same time, it upholds the epistemological contribution of "the languages," which *The Origin of German Tragic Drama* demonstrates through its literary rehabilitation of allegory, as a distinct and significant "area of truth" relinquished by mathematics as "true knowledge." For the confrontation between the Frankfurt School and the Vienna Circle, Benjamin's work on mathematics and language set up a key dialectical tension between two forms of thought, one expressed in mathematics that circumvents representation and the other mediated by language and representation.

Adorno's initial engagement with the logical positivists gave the tension between mathematics and other forms of knowledge the explicitly political dimension that we see in his and Horkheimer's confrontation with the Vienna Circle. In particular, contemporary philosophical movements including "the new Vienna School" and the "mathematical logicians" (*Logistiker*) such as Russell provided the backdrop against which Adorno proposed a new philosophy that stepped in amid the downfall of idealism, when thought could no longer claim to grasp "the totality of the real" (A 1:321 and 332).[33] Against the reconstruction of the question of being (*Dasein*) posed by

Heidegger and the "liquidation of philosophy" threatened by logical positivism, Adorno's "The Actuality of Philosophy" ("Die Aktualität der Philosophie," 1931) proposed a philosophical method of interpretation, which argued that the contradictions in the details of a text give insight into the social milieu that produced it.[34] Setting up his well-known polemics against Heidegger as well as attacking Carnap and Russell, the text depicts logical positivism and mathematical logic as dismantling and dissolving ("liquidating") all opposing modes of philosophical investigation. In his characterization of the logical positivists, Adorno's emphasis falls on the restrictiveness of their philosophical vision: speculative propositions that go beyond the "range of experience" must be found, Adorno writes, "solely in tautologies, in analytical statements." By forbidding anything that may go beyond what can be verified through experience, logical positivism turns philosophy, according to Adorno, "solely into an authority of order and control for the individual sciences [*Einzelwissenschaften*]" (332). This exclusion of any speculative, linguistic, and nonempirical judgments from philosophy in favor of the "tautologies" and "analytic statements" of mathematics framed the later dispute with logical positivism as a choice. Either philosophy could embrace Adorno's notion of interpretation or it could turn into the guidelines governing the real production of knowledge via mathematics in the "individual sciences"—that is, chemistry, physics, and biology, to name a few. If knowledge became nothing more than mathematics, philosophy and language would be expendable.

For Horkheimer and Adorno, however, the proposed exchange of the content and language of philosophy for mathematics would leave philosophy incomplete, unable to account for central aspects of thought, in particular, its own materiality. Adorno's missives to Horkheimer leading up to "The Latest Attack on Metaphysics" make this point with the help of Benjamin's distinction between the presence of representation in language ("proper meaning") and its alleged absence in mathematics ("improper meaning").[35] For Adorno, the attempt in mathematics to abandon meaning, the ability to signify something else, constitutes the philosophical flaw of the logical positivists' proposal to reduce thought to mathematics: "In principle, there's nothing behind this [perspective] but the fact that all logic, in truth, is removed from being and is not materially independent vis-à-vis being. To put it another way: the principle difference between language and mathematics, disputed by the mathematical logicians, proves itself here. In the strictest sense, logic cannot be mathematized—in the sense that 'consciousness,' that is, the categorical forms, still depend, even in this most formal sphere, on 'being'—that is, on their substrate."[36] This passage is confusing, but significant. Its main point is that mathematics, taken as the "formal sphere" of thought, is incongruent with the totality of thought. By detaching mathematics from "being," thought may feign immediacy. However, even as mathematics, "consciousness" depends on but is now unable to account for its material "substrate," such as the mechanisms of representation, writing, and communication. For Adorno, this incongruence was political, in that it not only limited philosophy but also threatened to detach thought from the real world—understanding and intervening in the problems of which, with the rise of National Socialism, was of the utmost importance for Horkheimer and Adorno. At the same time, the passage conflates all of "mathematics" with the area of mathematics studied by "mathematical logicians." Furthermore, although the passage again upholds the materiality of mathematics, it excludes the possibility for Horkheimer and Adorno that mathematics could function as a form of aesthetic or linguistic mediation, ideas I explore in subsequent chapters. Instead, both mathematical logic and mathematics transformed for Horkheimer and Adorno into a symbol of an incomplete philosophy, one that excludes language as the material instantiation of thought.

Horkheimer's public attack on logical positivism adopted Benjamin and Adorno's view that philosophy conducted via mathematics excluded language, but it also added salient new terms. For Adorno, the Vienna Circle liquidated forms of philosophy not expressible through mathematics, excluding questions of language, meaning, and representation. For Horkheimer, this exclusion also liquidated philosophical style. "Moreover, in their logic, these gentlemen disregard not only the relationship between word and meaning," Horkheimer responds to Adorno, "but, relatedly, also the connection of words and sentences to a stylistic unity."[37] If the program of the logical positivists consisted of eliminating normal language from science in favor of mathematics, then, for Horkheimer, they discarded the contribution done by the linguistic and aesthetic style of philosophy as the relation of not only *word* to *meaning* but also *word* to *sentence*. Indeed, at least on the surface, questions of style and presentation are anathema to mathematical formulae: $2 + 2 = 4$ in the same way that $4 = 2 + 2$. Horkheimer's criticism in "The Latest Attack on Metaphysics" thus focuses on mathematics in regard to this seeming disregard for language. The new logic advocated by the logical positivists, Horkheimer claims, "uses signs like mathematics for all formal elements as well as for the individual operations," meaning that "the linguistic elements are viewed without regard to their relationship to reality—that is, to the truth or untruth of the thought to which they belong" (H 4:144 and 143). Ultimately, Horkheimer's text focuses less on language than on how the logical positivists fetishize "singular elements of scientific activity," such as statements of scientific protocol, "making out of them a kind of religion."[38] Nonetheless, the threat that mathematics seemingly posed to language in logical positivism helped determine two central elements of Horkheimer and Adorno's emergent vision of critical theory, to which I turn in the remainder of this section.

The first element of Horkheimer and Adorno's thought brought into focus by the juxtaposition of mathematics and language is their critical image of mathematics as a mode of thought that excludes the concept of mediation. In the politics of academic philosophy in the 1930s, logical positivism's substitution of mathematics for language and representation as a medium of thought pointed to its immanent failure as an all-encompassing program for modern philosophy. As Horkheimer explains it in "The Latest Attack on Metaphysics": "The way in which the given is mediated here through thought, the way in which relations between objects are made visible, differentiated, and transformed, the linguistic structure, which effects the interaction [*Wechselwirkung*] of thought and experience, this inner development, is the mode of representation or the style. It is an insurmountable obstacle for this logic" (H 4:146). Logic in the form of mathematics sheds aesthetic style for formalism, impervious to "the mode of representation or the style." Again, such sentiments elide the idea that mathematics relies, despite the logical positivists' intentions, on its contents and the mechanism of representation. Instead, the conceptual association among mathematics, restrictiveness, and exclusion specifically in terms of mediation—coded here in terms of "insurmountability" and "difference"—would become a reoccurring exegetical trope for Horkheimer and Adorno's critical theory. As

Dialectic of Enlightenment claims: "For enlightenment, anything that does not reduce into numbers, finally to the one is illusion; modern positivism relegates it to poetry" (A 3:24). In the phase in which Horkheimer and Adorno worked to delineate critical theory from its philosophical competitors, mathematics signaled a powerful, but ultimately incomplete, form of thought that excluded language and representation—two concepts central to their emerging vision of philosophy.

This seeming indifference to language and representation in mathematics helped delineate a second element in Horkheimer and Adorno's vision of critical theory, namely its philosophical scope. Recall that, according to the logical positivists, those thoughts that cannot be translated into mathematical logic should, in Wittgenstein's words, "be passed over in silence." For Horkheimer and Adorno, this standpoint threatened to ignore the power that linguistic and artistic mediation—such as propaganda, political rhetoric, and new forms of media—commanded in the real, political world, especially amidst the rise of anti-Semitism in Europe during the 1930s. In contrast, as the "mathematical logicians attempt to translate language into logic (= to make it mute)," Adorno explains to Horkheimer. "I want to make logic speak [*die Logik zum Sprechen bringen*]."[39] Philosophy must explicate logic through both the content and the form of language, lending it, in effect, voice. Instead of exploring what mathematics could add to a theory of language, mathematics became for Horkheimer and Adorno symbolic of a theory devoid of language. As the antithesis of representation and mediation, as pure immediacy, mathematics delineated in negative terms "the whole aspiration of knowledge," which lay, as they put it in *Dialectic of Enlightenment*, not in the "mere perception, classification, and calculation" of knowledge, "but rather precisely in the determinative negation of whatever is immediate [*des je Unmittelbaren*]" (A 3:43).[40] As first envisioned by Horkheimer and Adorno, the "aspiration" of critical theory was thus an examination of the functions and effects of mediation ("the mediate") as critique of the status quo ("the immediate") that, through negation, determined the "positive" direction of thought. Accordingly, only by viewing form and content, being and thought in their dialectical "interaction" (*Wechselwirkung*) could one take account of the epistemological contribution and political consequences of aesthetic and linguistic mediation, instead of discarding them as poetry or myth. Only as dialectics and not mathematics, Horkheimer and Adorno believed, could thought hope to understand, let alone resist a social reality increasingly mediated by mass consumer culture and totalitarian politics across Europe in the 1930s.

Furthermore, this image of mathematical formalism as the philosophical exclusion of language continued to delineate the contours of Horkheimer and Adorno's vision of critical theory beyond *Dialectic of Enlightenment*. For instance, Adorno's text "The Essay as Form" ("Der Essay als Form," 1958) returns to and expands on the idea that logical positivism's emphasis on mathematics, as he put it in 1931, "dredges up the contours of all that in philosophy that is subject to a different authority than that of the logical and the individual sciences" (A 1:333). This authority, Adorno explains in the 1950s, is linguistic mediation; I quote, for effect, in full:

> The general positivist tendency which sets, through research, every possible object in firm opposition to the subject comes to a halt—as at all other times, so here too—when it begins to separate form and content: like trying to discuss the aesthetic unaesthetically, stripped of any similarity with its object, without turning into a philistine and a priori drifting away from that object. According to positivist custom, content once fixed in the primal image of the protocol sentence should not care about its presentation, which is conventional and not required by the matter itself; for the instincts of scientific purism, every impulse of expression endangers an objectivity that jumps out after the removal of the subject; such expression thus endangers the dignity of the material, which, so it is claimed, proves itself all the more the less it relies on form, although the principle of form lies in presenting the material purely and without addition. (A 11:11–12)

I will return to the notion of "protocol sentences" and the "removal of the subject" in the next section. What I want to call attention to here is how, even if the passage does not mention mathematics explicitly, it still reiterates conceptually Horkheimer and Adorno's objection that thought reduced to mathematics is synonymous with an incomplete vision of thought. It excludes the link between "form and content" and threatens "the dignity of the material." To think with mathematics means, in other words, to think "unaesthetically." In contrast, to think through the essay means, for Adorno, to "penetrate deeply into a matter" by holding systematic and antisystematic thought in dialectical suspension.[41] Indeed, this passage and the final tortuous sentence demonstrate that thinking that calls itself critical must take seriously the material codetermination of form and content not only as a constitutive feature of its objects of study but also as a presentational strategy. The final, winding thought, which constitutes a single sentence in the original German, suggests the deep linguistic contribution to critical thought, one that would be lost were thought to become mathematics. One task of critical theory, at least as Horkheimer and Adorno first saw it, lay in bringing the epistemological and political dimensions of thought seemingly foreclosed by mathematics—namely, that of language and mediation—to bear in philosophical practice.

At this point, we start to see images of language and critical thought that may appear more familiar to critical theorists in the present. The concept of language and philosophy constructed by Horkheimer and Adorno sits in explicit contradistinction to the mathematical logic employed by logical positivism. Here language is irreducible to its positivistic, communicative function—its ability to classify, calculate, and exchange knowledge. Instead, language provides the crucial medium between thought and being, and through its own material substrate, it constitutes a means of political intervention in terms of philosophical style.[42] Language and philosophy both depend on their content as much as on their form: the power of language lies in the dialectical interaction between what I say and how I say it.[43] Furthermore, as Horkheimer and Adorno's criticism of logical positivism subsumed mathematics together with mathematical logic, both also turned into a symbol of the task of their critical theory—in negative. As the very point of incongruence among logic, language, and reality, which critical theory seeks to expose, mathematics turned into a synecdoche for an exclusionary type of reasoning, one that fails to take account of the mediated nature of existence. To be sure, existence is not exhausted, as Horkheimer and Adorno claim, by the symbols and operations of mathematics. But what this concept of language and view of mathematics fails to take account of is the material dimension of mathematics that we see in negative mathematics: the linguistic and representational aspects that are indeed reducible but not erasable in mathematical thought. Even as mathematical logic reduces logic to mathematical signs, the traces of language and mediation remain, however austere and neutral these signs may appear. This side of mathematics bears the possibility for exploring the productive negativity that lies in representation, which remains closed in Horkheimer and Adorno's vision of mathematics.

Protocol Sentences: Experience without Subjectivity, or the Rise of Fascism

Horkheimer and Adorno's criticism of mathematics hinged on not only its alleged expulsion of language but also its exclusion of subjectivity from philosophy. Readers familiar with Horkheimer and Adorno's collaboration may think here of the dual moments of self-restriction and domination in bourgeois subjectivity that they see as the product of enlightenment rationality. In *Dialectic of Enlightenment*, for example, Odysseus represents the prototypical bourgeois subject, denying himself the fulfillment of the Siren's song and forbidding his men even its aesthetic gratification while extracting their labor (A 1:49–54).[44] When Benjamin and Adorno attended the Congress of the Unity of Science, such an understanding of subjectivity and the role that the subject played in the creation and maintenance of society would have appeared impossible in the logical positivists' intellectual framework. Indeed, as Adorno had advised Horkheimer, a critique of logical positivism had to address the idea of not only "token logic" but also "protocol statements," which led to a concept of "experience without the subject, that is, without the human."[45] For the logical positivists, such protocol sentences integrated empirical observations into the project of knowledge, framed in a purportedly neutral language that could be manipulated according to the rules of mathematics. For Horkheimer and Adorno, the role of "subjects" and "humans" as philosophical and political actors in the world disappeared when knowledge became the mere recording of protocol statements and operations of mathematics. Further subsuming mathematics in Horkheimer and Adorno's attack on logical positivism, this aspect of their critique turned mathematics into a mere instrument of reason akin to a form of political quietism that, in this initial version of critical theory, accepted and, hence, helped spread Fascism.

Before exploring the political ramifications that Horkheimer and Adorno associated with mathematics, it helps to understand the details of the logical positivists' theory of protocol sentences and its relationship to mathematics. According to the logical positivists, such protocol sentences supposedly offered a standardized and universally intelligible schema for knowledge, based, as in Carnap's initial definition, on the "original protocols" of scientific observations made by "a physicist or a psychologist."[46] Protocol sentences sought to minimize the ambiguities of conventional language, taking instead the form of the tautological statements of mathematics as formal vessels detached from content; on Carnap's example: "Arrangement of experiment: at such and such positions are objects of such and such kinds."[47] For the logical positivists, protocol sentences fulfilled the task previously assigned to the synthetic judgments: incorporating empirical data into knowledge and serving as experiential building blocks of knowledge that, like mathematical axioms, could be combined and unfolded according to "logical-mathematical rules of inference."[48] While members of the Vienna Circle debated the conceptual particulars of the theory and use of protocol sentences, the ultimate goal lay in creating a formalized language of knowledge: "all empirical statements can be expressed in a single language," Carnap claims, "all states of affairs are of one kind and are known by the same method."[49] As a mathematic-scientific replacement for the historical languages, such a protocol statements could provide the means, so the logical positivists hoped, for scientists to construct and communicate knowledge, seemingly free of the uncertainties of subjectivity and language.

As the idea of recording empirical data in protocol sentences and analyzing it with mathematics may suggest, an important difference between Horkheimer and Adorno and the logical positivists lay in how they approached the concept of experience. According to the logical positivists' *The Scientific Conception of the World*, experience separated the spurious claims of metaphysics and poetry from those of science (*Wissenschaft*) in logical positivism: metaphysical judgments "are meaningless [*sinnlos*], because [they are not] verifiable and without content. *Something is 'real' by virtue of being incorporated into the total structure of experience.*" For the logical positivists, "experience" meant "what is immediately given," what the world presented to perception, which protocol sentences recorded and mathematics ana-lyzed.[50] In contrast, experience for Horkheimer and Adorno was the com-plex product of social and psychological mediation, as discussed in the previous section in regard to language.[51] Take an example from Adorno's *Minima Moralia* that analyzes the state of domestic life: "Modern man wishes to sleep close to the ground like an animal, decreed a German magazine before Hitler with prophetic masochism, in order to abolish with the bed the threshold between waking and dreaming" (A 4:42). According to the passage, the modern experience of something as mundane as sleep is both a social product (mediated by "a German magazine") and a psychological product (mediated by masochism and the desire to return to nature). For Adorno, these forces turned sleep, far from "immediate," into a mechanism of the culture industry. Protocol sentences could record the decrease in bed height, but ignored the social, economic, and psychological forces that con-dition this decrease. While both perspectives attempted to incorporate ex-perience into philosophy, this inclusion for the logical positivists came in the form of an immediacy that they posited as primary, captured in a math-ematized language and studied according to mathematical rules. For the early critical theorists, thought must also take account of experience, but could not separate it from the social and subjective factors that codetermine what we experience as the seemingly immediate world.

As real as the threat may have been that the purported immediacy in logical positivism would preclude the subjective element from the concept of experience, Horkheimer and Adorno's critique of immediacy continued to amalgamate mathematics with their rejection of logical positivism. As Adorno writes to Horkheimer, the potential exclusion of the subject from experience pointed further to the incompleteness of logical positivism as a comprehensive philosophical program:

> The tendency toward resignation regarding the concept of experience asserts itself conclusively with the Neo-positivists. And in such a fashion that in order to get hold of experience as an absolute, as a tangible possession of science, they attempt to liberate the concept of experience completely from its inherent subjective moment—nothing different, that is, than from the awareness of human labor. Just as mathematical logic tries in vain by means of logic to manipulate away the relationship toward being (and, in the final

instance, societal being), the new theory of experience tries to dispose with the concept of the subject and, with it, any possibility of intervening in the world postulated as absolute facts. This is supported by all the concepts, like determination, the matter of fact, the protocol sentence, etc.—variants of the traditional positivistic concept of the given, treated naïve-realistically—with the purpose of eliminating the subject from experience and, thus, seemingly divesting experience of relativity, "securing" it. In truth, it serves instead to isolate the content of experience from anything human and all forms of human activity.[52]

Here and in the letter to Horkheimer that contains this extract, Adorno makes three moves essential to the image of mathematics established by Horkheimer and Adorno. The first position regards mathematics and mediation: although the passage points out that "mathematical logic" depends on being, it assumes the logical positivists' claim that mathematics is an immediate mode of thought. Instead of exploring how mathematics, too, mediates, Adorno's letter turns it into the very immediacy that he wishes to dispute. Drawing on their elision of subjectivity in favor of "pure being," the second of these points traces "the Neo-positivists" and mathematical logic back to the "origin of bourgeois thought" in the Enlightenment. They were the heirs to Francis Bacon's new empirical methods for the natural sciences and Thomas Hobbes's postulate of the "crude facts" of existence that necessitated a strong monarch.[53] In this history of thought, mathematics became, along with logical positivism, the latest embodiment of the Enlightenment, responsible, they later claimed, for the contemporary reversion of society back into barbarism.

The third position set up by Adorno's letter to Horkheimer concerns the politics of the logical positivist's "naïve-realistic" approach to experience via protocol sentences, which Adorno's letter compares to mathematics. By "naïve-realistic," Adorno meant the idea that there exists a "world postulated as absolute fact" that exists outside the subject and presents itself to the subject as immediate, "raw" experience. Recall that the logical positivists viewed "the world," as articulated by Wittgenstein, as "the totality of facts [*Tatsachen*]"; the world existed as a "totality" that offered itself "immediately" to perception, that the observer recorded in protocol sentences and the scientist analyzed according to mathematical-logical operations.[54] For

Adorno, this view comprehended experience only in terms of "first nature" (the so-called natural world), while excluding the Hegelian concept of "second nature" (the world created by human actions and by the human mind)—the effects of which a text such as *Minima Moralia* lays bare.[55] What is important here, however, is how this problematic view of reality lay for Adorno as much in the logical positivists' relegation of experience to "the determination, the matter of fact, the protocol sentence" as in the mathematization of logic. In the same way that mathematics feigns independence from the world, this notion of experience lacks "consciousness of human labor," "cutting out" the material contributions of subjects to experience and "isolating" experience from "humans and every type of human activity." Such restrictiveness points to the shortcomings of logical positivism but also opens a new dimension of Horkheimer and Adorno's criticism: eliminating humans from experience reveals logical positivism's impotence as a mode of thought that forfeits "any possibility of intervening in the world." Instead, thought records and interprets a given "totality of facts" via the seemingly neutral signs of mathematics while withholding judgements about these facts. For Horkheimer and Adorno, protocol sentences and mathematics formed two sides of the same coin: they naively assumed that the world is an unalterable "totality," but also inhibited thought from grasping its own contribution to and ability to intervene in the world.

At this point in Horkheimer and Adorno's criticism of logical positivism, their accompanying criticism of mathematics began to assume distinct political contours as a sign of complacency in the ever-worsening political situation of the late 1930s and 1940s. As we have seen, the political dimension lay for Adorno in the incompleteness in terms of language and subjectivity that the logical positivists' turn to mathematics meant as an all-encompassing philosophical platform. What was politically troublesome about logical positivism and, by association, mathematics was "the elimination of the concept of the subject" through protocol sentences' alleged "rigor" and "neutrality." On the contrary, Horkheimer explains:

> The identification of this abstract moment of exactitude, which upon closer inspection turns out to be its opposite, with the concept of truth as such is only the transfiguration of the silence of these latest liberals. With it, they help sanction and spread the horror that through their totalitarian successors has

come over the world. This contemporary philosophy does mathematics, the rest is silence. Such abstinence is by no means passivity, since it indeed makes abundantly audible propaganda for itself. Much more, it represents a part of the cultural apparatus whose function it is to make humanity mute.[56]

The letter's reference to Hamlet's last words ("the rest is silence") is unequivocal: "doing mathematics" at the total cost of language ends in tragedy and death.[57] At this point, Horkheimer and Adorno explicitly implicated mathematics in the "horror" of Nazism that had stripped Jews in Germany of their rights and citizenship, had sent members of the Frankfurt School and Vienna Circle into exile, and in two years would start World War II. In particular, the passage designates mathematics as a refuge for political "abstinence" that, instead of advocating philosophy as a means of political intervention, fell "silent" in the face of the collapse of liberal democracy in Germany and the rise of its "totalitarian successors" in 1933. For Horkheimer and Adorno, the turn to mathematics made "humanity mute" by demanding that we pass over, as Wittgenstein says, "in silence" modes of knowledge that we cannot express in protocol sentences and manipulate in the signs and with the operations of mathematics—such as economic and political critiques of totalitarianism. This persistent substitution characterizes much of Horkheimer and Adorno's criticism of mathematics: it is not the political orientations of individual logical positivists, but rather their focus on mathematics at the cost of philosophical modes of political intervention that brings them and mathematics into the orbit of Fascism.

For Horkheimer, the exclusion of the subject from experience evinced complicity with authoritarianism, but replacing thought with mathematics also transformed thought into an uncritical instrument. As mathematics and protocol sentences made language and the cardinal questions of philosophy irrelevant, it also, Horkheimer added in "The Latest Attack on Metaphysics," threatened the critical and historical capacity of thought:

> The thought—not only, for example, in physics, but rather in knowledge as such—to allow the subject to disappear by grasping individual differences themselves as a series of facts is itself a research principle to be circumscribed with caution. The transformation of this postulate into the belief that it is fundamentally applicable at any given historical movement leads necessarily

to an unhistorical and uncritical concept of knowledge and the hypostatization of the individual natural-scientific procedures. (H 4:123)

This criticism of protocol statements manipulated with mathematical operation adapts and expands Adorno's criticism of mathematical logic: "hypostatizing" knowledge as "series of facts" exchanges subjectivity for mere factuality. Knowledge expressed exclusively as mathematical statements forfeits thought's ability to understand and criticize the world not as a necessary absolute fact, but as the construction of human history. Moreover, mathematics not only cannot grasp but also perpetuates reification, becoming here a universal method applicable "at any given historical moment." In contrast to dialectics (and "The Latest Attack on Metaphysics" readily point out its advantages), a philosophy that excludes subjectivity from experience turns thought into a simple tool to arrest immediacy into the tautological forms of mathematics. So construed, philosophy would lose the ability to grasp the historical conditions that led to the contemporary situation in Europe in the 1930s, let alone critically evaluate and resist its consequences.

The eschewal of historical and critical thought brought on by the exclusion of subjectivity turned the philosophical claim against mathematics political, which Horkheimer's "The Latest Attack on Metaphysics" made public and his and Adorno's later work on critical theory continued. If we restrict philosophy to mathematics, physics to an aggregate of seemingly factual protocol statements, and language to its mere communicative function, then little remains for thought to counter political oppression. "For large swaths of the middle classes who have fallen behind in the free play of economic powers, there remains," the text claims, "where they have not fully attached themselves to the economically most powerful, only the possibility of a silent existence [*die Möglichkeit einer stillen Existenz*], restraint in all decisive questions. Thought relinquishes its claim to both be critical and set goals" (H 4:153). What Horkheimer fears is that if thought becomes mathematics, individuals ("the middle classes") will lose their ability to address "all decisive questions"—that is, to participate in society, be "critical" of it, and "set goals" in hopes of changing it. In the 1930s, knowledge as mathematics was a politics of silence, reflected in the dismantling of language and loss of the "critical" capacity to grasp the contemporary political situation as the product of human action. In this regard, the fact that individual members of the Vienna Circle were opponents and victims of totalitarianism mattered little to Horkheimer; the exclusion of the subject from knowledge in favor of mathematics made logical positivism, as a whole, politically complicit in it.[58] "If science takes on such a character, if thought per se loses the willfulness and imperturbability to penetrate a forest of observations," Horkheimer explains in reference to the mathematician and logical positivist Hans Hahn (himself of Jewish decent), "then they passively take part in universal injustice" (127). According to this view of critical theory, protocol sentences and mathematics worked not in the interest of political emancipation and human freedom, but rather were tools of repression and subjugation.

The political implications of the disappearance of subjectivity from experience that Horkheimer and Adorno associate with mathematics point forward to their critique of instrumental reason. Instrumental reason refers to modes of thought that take the means of thought as its ends.[59] For instance, to think via protocol sentences and according to the rules of mathematics, as suggested by the logical positivists, substituted the real goal of reason—to create a self-reflective society—with a system of knowledge built out of the formalized, content-less statements of mathematics. In later texts such as *Dialectic of Enlightenment* and Horkheimer's *Eclipse of Reason* (1947), instrumental reason targets efficiency and profit, rather than, say, justice or social emancipation. Already in the 1930s, logical positivism along with mathematics worked along these lines for Horkheimer, not toward a historical and critical understanding of contemporary society, but instead as "modest servants of industry," benefiting from and propagating further "the pre-established harmony of specialized science and barbarism [*Fachwissenschaft und Barberei*]."[60] As he puts it in "The Latest Attack on Metaphysics": "This ideology—the identification of thought with the specialized sciences—works in the face of the leading economic forces, which avails itself of science and all of society for its specific purposes, in effect toward the perpetuation of the contemporary situation" (H 4:154). As the method unifying the "specialized sciences," mathematics can expedite thought and standardize communication, but expedition and standardization only intensify the status quo, instead of qualitatively changing it. During World War II, this stubborn adherence to and repetition of the given, recorded in protocol statements and manipulated like mathematics, turned in *Dialectic of Enlightenment* into the driving force behind reason's regression to barbarism.

Even after the war, texts such as Horkheimer's *The Eclipse of Reason* drew on and disseminated the connection between mathematics and instrumental reason. In *The Eclipse of Reason*, mathematics exemplifies the Janus-face of instrumental reason: it expands knowledge, but also hampers the individual's ability to resist mass manipulation, leading, ultimately, to dehumanization. For Horkheimer, the meaning of *reason* had undergone a radical transformation in advanced industrial society. When I receive a parking ticket, for instance, I ask whether I complied with the pertinent regulations, instead of inquiring into the justness or unjustness of these laws. Here, Horkheimer's argument revisits and expands his association of neo-positivistic thinking and mathematics from the 1930s.[61] "Complicated logical operations are carried out without actual performance of all the intellectual acts upon which the mathematical and logical symbols are based," Horkheimer explains. "Such mechanization is indeed essential to the expansion of industry, but if it becomes the characteristic feature of mind, if reason is instrumentalized, it takes on a kind of objectivity and blindness."[62] As in the 1930s, mathematical and logical symbols signify in *The Eclipse of Reason* the shift from a concept of reason, as the potential basis for progressive society, into a "blind" form of subjective reason focused exclusively on self-preservation and self-advancement—here, in economics. Focusing more on self-preservation than on the reasonable constitution of society predisposes instrumental reason, Horkheimer claims, to "tilt over into fascism."[63] This criticism was hyperbolic, but salient for their vision of critical theory: mathematics as the sole ends of philosophy may prove economically advantageous to the subject, but it says nothing about the society—its justness or unjustness, humanity or inhumanity—in which that subject lives. As discussed in the following section, these contradictions provided, according to Horkheimer and Adorno, the intellectual cognitive conditions that made Auschwitz possible and that critical theory seeks to expose.

The way Horkheimer and Adorno associated mathematics with instrumental reason and the politics of domination and control helped determine the direction of the critical project after *Dialectic of Enlightenment* and *The Eclipse of Reason*. For instance, critical theory still holds contemporary relevance, as Andrew Feenberg shows, less in actualizing revolutionary activity than in providing modes of "noninstrumental practice" to transform the horizon of meaning in which such activity takes place. Without resorting to the "manipulation and control" of instrumental reason, critical theory expands cultural horizons of meaning, enabling political action that works toward the inclusion of new forms of social existence and against exclusion and repression.[64] I agree with this assessment, but caution that if critical theory too readily accepts the way Horkheimer and Adorno associated mathematics with instrumental reason and the politics of domination, it risks giving up the critical potential of mathematics and any other interpretive tool that Horkheimer and Adorno broadly associated with instrumental reason, such as technology. As explored in subsequent chapters, negative mathematics allows us to view mathematics not only as a potential tool of calculation and equation but also as a powerful mode of aesthetic signification and cultural analysis. Moreover, negative mathematics bears the possibility of employing mathematics as a limited instrument that offers the basis for more inclusive theories of knowledge and history. The problem with the overdetermined equation of mathematics with instrumental reason and its dismissal as such is that it surrenders mathematics—not to mention technology—to the forces of industry, government, and, ultimately, the opponents of critical theory.

Logic and Empiricism: From Logical Positivism to Metaphysics, from Enlightenment to Myth

Building on its exclusion of subjectivity and language, Horkheimer and Adorno's criticism of mathematics hinged on what they interpreted as the metaphysical dimension latent in the logical positivists' equation of mathematics and thought. Making this point was the conscious and contentious intervention of Horkheimer's essay "The Latest Attack on Metaphysics," because metaphysics—speculative thought that lacks empirical grounding— was the very thing that the logical positivists wanted to eliminate from knowledge. The main thesis of Horkheimer's essay holds that logical positivism as an epistemological program represented a dialectical return to

metaphysics as naïve and sectarian as the "romantic spiritualism" of Wag-
ner, the *Lebensphilosophie* of Bergson, and the "material and existential phenomenology" of Heidegger (H 4:112). For Horkheimer, the totalizing view
of mathematics in logical positivism facilitated this return, revealing the
deep contradiction in bourgeois thinking that made Fascism possible. "Metaphysical illusions and higher mathematics," he writes, "constitute in equal
measure elements of its [i.e. bourgeois society's] mentality" (111). To be sure,
this was not a defense of metaphysics vis-à-vis mathematics, but rather the
dialectical move that would underpin *Dialectic of Enlightenment*: equating
empiricism and mathematics with the totality of knowledge was itself a
nonempirical, metaphysical statement. When upheld as the sole form of
knowledge, mathematics thus exemplified not only how logical positivism
entailed another form of metaphysics, but also how reason turns into the
mythic mode of thought from which it had once promised liberation. The
transformation of thought into mathematics, into its seemingly ritualistic
symbols and repetitive operations, was Horkheimer and Adorno's final
charge against mathematics, which incorporated it into their master narrative of the demise of enlightenment itself.

In "The Latest Attack on Metaphysics," mathematics facilitates the logical positivists' surreptitious return to metaphysics because it constitutes one
side of their dual epistemological emphasis, expressed perhaps best by the
movement's other name, logical empiricism. "The unification of empiricism
with modern mathematical logic," as Horkheimer describes the Vienna
Circle, "is the essence of this newest school of positivists" (H 4:116). Their
ideal of knowledge is thus empirical facts manipulated as if they belonged
to a system of mathematics: "knowledge as a mathematically formulated unified science deduced from the fewest possible axioms" (114). In terms of the
type of knowledge privileged by logical positivism, such "unification" aptly
characterizes what the logical positivists' *The Scientific Conception of the World*
acknowledges as meaningful statements that count as knowledge: statements
whose "meaning can be determined by logical analysis or, more precisely,
through reduction to the simplest statements about the empirically given."[65]
These are the seemingly immediate observations recorded in protocol language: subject X mixed chemical solution Y at time Z—the convergence of
terms developing throughout this chapter. The logical positivists envisioned

knowledge to be the combination of tautological statements about the immediate, "empirically given" and the "logical analysis" of these statements via the symbols and operations of mathematics. Recall, too, that such knowledge should leave little room for philosophical interpretation; according to Carnap, "facing the implacable judgment of the new logic," the "old" logics from Plato to Kant and Hegel prove to be "not merely simply false with regard to their content, but also logically untenable and, therefore, meaningless [*sinnlos*]."[66] For the first generation of critical theorists, however, the irrationality and anti-Semitism of the Nazi regime and the ever-worsening situation in Europe indicated not the unification, but rather the radical divergence of logic and reality.

For Horkheimer, the totalized turn to mathematics pinpointed the incongruence of the two epistemological arenas to which logical positivism laid claim. Unifying logic and reality, as the logical positivists hastened to point out, had been the Enlightenment dream since Leibniz's *mathesis universalis*, which initially tried "to master reality through a greater precision of concepts and inferential processes, and to obtain this precision by means of a symbolism fashioned after mathematics."[67] The proposed unification of "reality" and "mathematics" represented the moment at which the modern iteration of this dream, logical positivism, faltered: "it becomes apparent that both elements of logical empiricism are only outwardly unified" (H 4:147). Mathematics' exclusivity and mismatch with reality, pointed out by Adorno, neglected matters of representation and subjectivity, "shamelessly" scarifying, Horkheimer claims, "all spheres of culture totally to irrationalism."[68] There is no doubt: these exclusions would have aggravated Horkheimer because they excised from epistemology the analysis of culture such as art and film, perhaps accounting for the essay's polemic tone.[69] But for Horkheimer and Adorno, this view of philosophy would have also been politically suspect in as much as it made a critique of capitalism and bourgeois society impossible. The more mathematics represented immediacy for Horkheimer and Adorno, the more it threatened to exclude from the purview of knowledge crucial phenomena such as language, mediation, reification, and the historical development of society as rationalization and class struggle. The idea that reality was only simple, mathematized facts was the worldview that first suggested logical positivism's link to conservative revivals of "neoromantic philosophy": "Both philosophical currents conceive of reality not in conscious connection with a particular historical activity as the epitome of tendencies, but rather hold themselves to reality in its contemporary form" (157). Holding thought "to reality in its contemporary form" was as reactionary in the 1930s as it is today, but it also equated the portion of reality accessible to mathematics with reality as a whole, which was itself a nonempirical, metaphysical claim.

In its focus on the "irrational" content incongruent with mathematical forms and excluded by logical positivism, Horkheimer and Adorno's criticism of mathematics drew on an argument central to Lukács's essay on reification in *History and Class Consciousness*, published in 1923. Reification is the process by which qualitative relationships between people turn into seemingly objective things, giving the institutions that shape social life an air of neutrality and immutability. The problem with reification is that its products are not flawless, but rather human creations, some of which harbor contradictions that Lukács called the "antimonies of bourgeois thought."[70] The first of these antinomies held that a fundamental incongruence persists in philosophy since Kant between the rational forms of thought and the irrational contents of modern life. On the one hand, "pure and applied mathematics have constantly been held up" by Kant and the Neo-Kantians, "as the methodological model and guide of modern philosophy. For the way in which their axioms are related to the partial systems and results deduced from them corresponds to the postulate . . . that every given aspect of the system should be capable of being deduced from its basic principle."[71] On the other hand, Lukács argues, this system of rationalism repeatedly comes up short in giving an adequate explanation of the totality of empirical objects and experiences that should be, theoretically, deducible in these "systems." Instead, rationality relegates their true nature to an unknowable and ungraspable concept, like Kant's "thing-in-itself." Bourgeois thought thus finds itself making a contradictory claim, demanding a system of universal rationality and, simultaneously, realizing "that such a demand is incapable of fulfillment" because of the persistence of such irrational elements.[72] Written in the wake of the Russian revolution, this contradiction signified to Lukács one of the antimonies that conditions the crises of bourgeois rationality, which, ultimately, would anticipate the economic transformation of society as a whole.

Horkheimer and Adorno's criticism of mathematics hinged on Lukács's notion of the incongruence of mathematical form and societal content, but also updated it for the era of Fascism. The inability to reconcile experience with the forms of logic and thought did not mean for Horkheimer and Adorno that we should abandon logic and thinking as a whole. Rather, as Adorno explains to Horkheimer, it reveals the impossibility of the attempt to bridge the "acknowledged dualism between mathematical logic and empiricism that runs through scientific philosophy as a whole."[73] In "The Latest Attack on Metaphysics," the mathematical-mechanical treatment of logic likewise points to the limits of logical positivism:

> Knowledge relates solely to what is and its reoccurrence [*das, was ist, und seine Wiederholung*]. New forms of being, especially those that arise from the historical activity of humans, lie beyond the theory of empiricists. Thoughts that cannot be subsumed fully in the prevailing forms of consciousness, but rather must be grasped as the aims and decision of the individual—any historical tendency that, in short, reaches beyond what is present and recurrent—do not belong, according to this conception, amongst the concepts of science. (H 4:120)

This passage does not mention protocol sentences and mathematical logic. Instead, it implies their complicity in maintaining "the prevailing forms of consciousness" (i.e., authoritarianism) by the sense of exclusion, reformulated here in terms of "what is" (the immediacy of the empirically given) and its "recurrence" (its repetitious manipulation through mathematical logic). This shift marks a conceptual development for Horkheimer and Adorno in which mathematics comes to mean the exclusion of not just language and subjectivity but also "new forms of being" through its focus on immediacy, repetitiveness, and reoccurrence. As I have shown, reality for Horkheimer and Adorno was neither immediate nor repetitive, but rather deeply mediated by both language and subjectivity. If we hold knowledge to the mathematical formulation of what is, then we restrict "what is"; conversely, if we present the analytic machine of mathematics with a fuller version of reality such as "the historical activity of humans," then it excludes, represses, and eliminates portions of that reality. For Lukács, the illusory promise of mathematics upheld as epistemology signaled a coming crisis in bourgeois society. A decade later, it indicated the manifestation of a contradiction already very much at work in bourgeois thought.

The impossibility of uniting mathematical logic and empirical knowledge of reality informed Horkheimer's accusation that logical positivism relapses into metaphysics, but it also set the conceptual tone for the image of mathematics that emerged in *Dialectic of Enlightenment.* Through its critique of the logical positivists' equation of mathematics and thought, Horkheimer's "The Latest Attack on Metaphysics" demonstrates how logical positivism, which he and Adorno associated with Enlightenment thinkers such as Bacon, dialectically transforms back into its opposite. Remember that the Vienna Circle at all points resisted metaphysics, in both philosophy and politics. The incongruence of mathematics and the empirical world, however, suggested to Horkheimer the metaphysical dimension to logical positivism's main epistemological claim. "Even when metaphysics is wrong to console humans with a being that fundamentally cannot be determined through the means of science," Horkheimer claims, "so too science becomes itself naively metaphysical when it mistakes itself for knowledge and theory as such and wishes to discredit the name of philosophy—that is, every critical attitude toward science" (H 4:158). The equation of mathematics and protocol sentences with knowledge as a whole is metaphysical, because it not only represents a speculative judgment about the constitution of "knowledge and theory" (namely, that they are, in the end, scientific) but also remains impossible to realize in reality. For Horkheimer and Adorno's critical theory, mathematics became a symbol for the contradiction of logical positivism, returning it to the type of thinking it hoped to vanquish, metaphysics.

In "The Latest Attack on Metaphysics," this sense of exclusion and contradiction that the text associates with mathematics indicates logical positivism's epistemological shortcoming as well as its reactionary politics of knowledge. For Horkheimer, the way that logical positivists' turn thought into mathematics was cultish. Already in his letters to Adorno, Horkheimer's criticism depicts logical positivism as epistemological sectarianism, referring to the logical positivists' "sectarian customs" of congresses (such as the one attended by Benjamin and Adorno in 1937), unified science, and unified language. They turned the "lean phrases" of mathematics "into a panacea in truly obsessive-compulsive manner."[74] Accordingly, logical positivism's restrictive vision of knowledge as mathematics failed to account for domains of the empirical, but it thus also takes on the shape of a metaphysical, religious belief: "This positive relationship to science does not mean, however, that their language is now, as they say, the one authentic and true form of knowledge. In comparison to the level of knowledge attainable today, the portion of reality encompassed by the disciplines is limited by not only the range but also the mode in which we speak about it. Just as it is incorrect to run counter to the results of science, so too is it naïve and sectarian to think and speak only in the language of science" (H 4:158). We recognize Horkheimer's dialectical juxtaposition of metaphysics and "the language of science" again, but here the passage emphasizes terms of "sectarian" restriction and self-containment. As the "language" of tautologies, mathematics embodied "the singular and true form of knowledge," making logical positivism an "self-enclosed *Weltanschauung*" more akin to the "worldview" offered by religion than to "science" (159). According to "The Latest Attack on Metaphysics," the Vienna Circle represented not a progressive scientific movement, but rather a "philosophical sect." Mathematics was, then, the cult language, the totemic chant, of this Enlightenment sect that turned experience into mere immediacy and thought into its blind repetition.

The idea that the logical positivists' investment in mathematics would turn their intellectual program back into metaphysics, its sworn enemy, marked the evolution of the philosophical concept that Lukács called the "antinomies of bourgeois thought." "The Latest Attack on Metaphysics" transformed the nature of this antimony from an economic contradiction into a contradiction immanent in reason itself.[75] In the same way that logical positivists' turn to the new mathematical logic of Russell and Whitehead eliminated metaphysics, it also sought to eliminate contradiction in favor of "clarity," overcoming, as stated in *The Scientific Conception of the World*, "logical contradictions, 'antinomies,' which pointed to essential mistakes in the foundations of traditional logic."[76] For Horkheimer, however, this claim to historicity ("new" versus "traditional logic") was at odds with mathematics itself: "By simply contrasting ideas it has chosen to consider as genuine and true to the achievements of thought that played and continue to play a role in human history, their logic falls completely outside the role of tautology and reveals itself as a subjective position totally opposed to empiricism" (H 4:147). There are really two contradictions at play here. The claim that mathematics is the "genuine and true" form of knowledge is itself a "subjective" and not "tautological" judgment. Furthermore, the passage contradicts the logical positivists' purported commitment to empiricism, which ignores "achievements of thought" that constitute the empirical world—such as Marxism, art, or literature. Mathematics, in other words, brought a contradiction to surface that indicated the nonviability of logical positivism as an epistemological program. It also, as Horkheimer explains in later essays such as "Traditional and Critical Theory," points to the "illusion" and "contradiction" (and not "neatness" and "clarity") proper to thought itself, exemplifying "something dark, unconscious, and opaque" that "reflects precisely the contradiction-filled form of human activity in modernity."[77] For Horkheimer and Adorno, mathematics constituted the opposing side in a debate over reason itself, whose darkest depths had taken center stage in Europe by the late 1930s and whose innermost contradictions it was now incumbent on philosophy to understand.

More than just rationalization and reification, this was the fuller image of mathematics' perniciousness that informed and was radicalized by *Dialectic of Enlightenment*. As the text famously argues, rather than realizing the emancipation and freedom promised by the historical period known as the Enlightenment and by *enlightenment* as the faculty of reason, rational thought already contains the seed of its return to mythic and barbaric forms of thought manifest in the rise of National Socialism and the horror of the Holocaust. While many have remarked on the totalizing, even nihilistic nature of Horkheimer and Adorno's historical narrative, I believe it is imperative that we recognize how this narrative also enshrines their conflation of mathematics and logical positivism, which in *Dialectic of Enlightenment* figures as the modern incarnation of the antique and early-modern forms of enlightenment—Odysseus and Francis Bacon.[78] "Positivism" assumes "the judicial office of enlightened reason" (A 3:42). In essence, *Dialectic of Enlightenment* translates Horkheimer's interpretation of logical positivism's return to metaphysics into the most recent example of reason's relapse into myth.

Generalizing the terms of the conflict in the 1930s, the notion of mathematics presented by *Dialectic of Enlightenment* sees in it the illusionary promise of enlightenment: "In the preemptive identification of the thoroughly mathematized world with truth, enlightenment believes itself to be safe from the return of the mythical," Horkheimer and Adorno write; enlightenment "equates thought with mathematics" (41). In a work that helped Horkheimer and Adorno establish critical theory, mathematics came to represent a modern paradigm of enlightenment thinking, which promised emancipation from myth and which aided in the quest of thought to preserve itself and dominate nature.

According to Horkheimer and Adorno, the incongruence between mathematics and the empirical world that paved the way for logical positivism's return to metaphysics also facilitated enlightenment's return to myth. In *Dialectic of Enlightenment*, the term *myth* refers to a prescientific mode of knowledge, characterized by the immediacy of the spirit world, repetition and ritual, and a belief in the animate power of objects. In its modern incarnation, mathematics and the exclusionary nature of the logical positivists' equation of mathematics and thought suggests that enlightenment entails as much emancipation as mythic restriction and control. "Enlightenment is mythic fear radicalized. The pure immanence of positivism, enlightenment's final product, is nothing other than a kind of universal taboo" (A 3:32). This fear-driven taboo not only excludes critical, historical thought but also condemns thought to repeat, through mathematical operations, whatever experience gives to thought, as if mathematics were a "ritual of thought": "Instead, mathematical formalism, however, whose medium is number, the abstract form of immediacy, holds to thought at pure immediacy. The factual is proved right, knowledge limits itself to its recurrence, thought turns into a mere tautology. The more the machinery of thought subjugates being, the more blindly it satisfies itself with reproducing it. Thus, enlightenment regresses to the mythology, which it never knew how to escape" (44). This text repeats the vocabulary of ideas that almost a decade earlier criticized the restrictive nature of "mathematical logic." As Adorno suggested, mathematics mediates, but, according to this passage, mediates poorly: it limits thought to the subject-less "immediacy" of experience, privileges "the factual" so that thought becomes a "mere tautology," and restricts knowledge to "recurrence." In "The Latest Attack on Metaphysics," these terms pointed to logical positivism's complicity in authoritarianism. *Dialectic of Enlightenment* reinterprets these terms as the totemic immediacy and repetition akin to the stories of "mythology." This is a key point: restricting thought to mathematical immediacy, tautology, and repetition results in a society and a concept of knowledge organized around mathematics, not reason. Moreover, it restricts or "subjugates" being to that which fits into the "machinery" of mathematics. In Horkheimer and Adorno's critical theory, this return to the terms of myth was a central intellectual gesture that found in thought reduced to mathematics not the liberation, freedom, and unity once promised by the Enlightenment, but instead the domination, restriction, and discord of Nazism.

In the middle of World War II, in other words, Horkheimer and Adorno transformed the logical positivists' equation of mathematics and thought into a core failing of reason as such. In *Dialectic of Enlightenment*, mathematics, when taken as the sole medium of thought, provided the conditions leading to the downfall of the Enlightenment project.[79] In its construction of this history of thought, Horkheimer and Adorno's collaborative magnum opus further subsumed mathematics as a foundational part of the dialectical return of reason to barbarism, but not only, as Habermas suggests, as the type of instrumental reason that critical theorists associate with rationalization and technological domination.[80] *Dialectic of Enlightenment* also sees in the logical positivists' failed attempt to unite mathematics and the empirical world the neurosis of reason. One of the text's central contentions is that thought represses its origin as nature: thought is part of nature, but its equation with mathematics holds it up in contrast to and as distinct from nature. While thought thus attempts to control nature in the name of self-preservation, the release of those parts of nature that it suppressed—excluded as "meaningless" when thought became mathematics—return in the form of the barbarism and violence of the twentieth century. The incongruence between mathematics and empiricism at the heart of logical positivism thus reveals contradiction rather than harmony at the heart of reason—the unbridgeable "break of subject and object" that serves as "the index of the untruth both of itself and of truth" (A 3:43). The ultimate irreconcilability of the subject (mathematics) and object (the empirical world) immanent in logical positivism underscored the contradiction of reason, which, for Horkheimer and Adorno, it then became the task of critical theory to expose.

Between Horkheimer's "The Latest Attack on Metaphysics" and *Dialectic of Enlightenment*, the idea in logical positivism of equating mathematics with thought came to play a pivotal role in an apocalyptic narrative of reason that called into question the integrity of reason, of enlightenment itself. "The absurdity of the situation, in which the violence of the system grows over humans with every step that releases them from the violence of nature," Horkheimer and Adorno conclude, "denounces the reason of the reasonable society as obsolete" (A1: 43). With the publication of *Dialectic of Enlightenment* in 1944, mathematics became the avant-garde of the same reason that enabled the Holocaust. In this phase of Horkheimer and Adorno's collaborative efforts to establish and disseminate critical theory, mathematics provided another name for the taboo of art, enabler of political authoritarianism, and a metaphor for the false promise, if not destructive pathology of enlightenment itself. The publication of *Dialectic of Enlightenment* thus solidified an interpretative posture toward mathematics that made it taboo for critical theorists who wished to follow in the tradition that the Horkheimer and Adorno helped establish.

As much as Horkheimer and Adorno's criticism of mathematics was a political critique of logical positivism, the association of mathematics with the demise of enlightenment in this initial vision of critical theory was also a politics of thought, in which mathematics symbolized the control and domination wielded by bourgeois society. This opposition was necessary, as Albrecht Wellmer explains, for critical theory to intervene in society, lest it remain in the abstract realm of pure reason. Already in Horkheimer's attack on logical positivism in the 1930s, "theoretical criticism had simultaneously and distinctly to be conceived as a part of critical *praxis*: in the last resort, theoretical disputes with bourgeois science could be maintained *in practice* only as a form of the class struggle."[81] The task of critical theory included not only naming how mathematics and other forms of natural-scientific inquiry aid domination and oppression in politics and the economy, but also resisting the mathematization of thought and instrumental reason as a whole. And yet, as Wellmer notes, "the philosophical intensification of the criticism of capitalism" in *Dialectic of Enlightenment* threatened to exclude the "scientific" (*wissenschaftliche*) basis of criticism of societies in which science itself had become a constitutive factor of modern life.[82] Later critical theorists, such as Wellmer and Habermas, worked to reintegrate the scientific element into a critical theory of society, viewing the normative claims of modern societies as intelligible only through a combination of philosophy and empirical social research.[83] While this postwar refashioning and expansion of critical theory may have brought critical theory back into dialogue with science, it was by no means a rehabilitation of the critical potential of mathematics.

Instrumental Reason: The Afterlife of Mathematics in Critical Theory

In this context, it is easy to see how mathematics has fallen out of the canon of critical theory after Horkheimer and Adorno. This section explores why the critical potential of mathematics remained on the margins of the critical project, even after its inaugural phase exemplified by *Dialectic of Enlightenment*. The mistrust of mathematics that emerged between Benjamin's work on language and mathematics and Horkheimer and Adorno's magnum opus set up an intellectual hostility toward mathematics that viewed it as the exclusion of language from philosophy, the erasure of subjectivity from experience, and the dialectical return to metaphysics and myth. Remarkable about the story of mathematics in Horkheimer and Adorno's conception of critical theory is its meticulous construction: what was a conflict between competing philosophical circles and schools became a foundational narrative in critical theory; what was once indignation over the exclusiveness of turning thought into mathematics became the pathology of reason that ended in the Holocaust.[84] While it may be tempting to dismiss these polemics as the excesses of, as Alexander Düttmann puts it, "the philosophy of exaggeration," we have seen throughout this chapter how this narrative obscured the critical potential of mathematics in Horkheimer and Adorno's own writings and helped establish what would (and would not) count as critical theories of language, aesthetics, and society.[85] Beyond Horkheimer and Adorno, the image of mathematics as a philosophy of exclusion and the antithesis to critical thought persisted in later formulations of critical theory, further foreclosing the potential contributions of mathematics to the critical tradition practiced across the humanities today.

After Horkheimer and Adorno's confrontation with the logical positivists, mathematics endured as a symbol of the type of thinking that critical theory took as its task to oppose. In the 1930s and 1940s, this initial vision of critical theory explicitly countered what in 1937 Horkheimer called "traditional theory," which strove for "a pure mathematical system of signs" (H 4:164). Such opposition also informed Herbert Marcuse's *One-Dimensional Man* (1964), which argued that the economic and technical needs of advanced capitalist society determined not only the skills and attitudes of its members but also their individual desires and aspirations.[86] Marcuse's text adapts Horkheimer and Adorno's objection that the mathematical formalism of neo-positivism remains "indifferent toward its objects." It also holds "contemporary mathematical and symbolic logic" in "radical opposition to dialectical logic." Formal logic from Aristotle to "new formal logic express the same mode of thought," according to Marcuse; "it is purged from that 'negative' which loomed so large at the origins of logic and of philosophic thought—the experience of the denying, deceptive, falsifying power of the established reality."[87] Thought idealized into mathematics eliminates thought's subversive potential as protest and refusal by excluding the historical and mediated dimension of reality "established" by human action. In contrast, the dialectical logic of critical theory "militates against quantification and mathematization."[88] As it had for Horkheimer and Adorno in the 1930s, critical theory for Marcuse counteracted mathematics and offered a type of thought that attended to its objects as a means of exposing and transcending the contradictions of a society rationalized by formal, mathematical logic.

Furthermore, for both Adorno and Habermas, mathematics continued to demarcate the disciplinary boundaries around critical theory on the intellectual map of postwar West Germany. Benjamin and Adorno's trip to the Congress of the Unity of Science in 1937 thus serves an apt prelude to the dispute over empirical sociology that emerged at the 1961 Conference of the German Sociological Society, which reflected and intensified the intellectual rivalries among critical theory, logical positivism, and mathematics that we saw forged in this chapter. Just over a decade after the Frankfurt School had officially returned to West Germany, Adorno and Habermas used the event and subsequent publications, known collectively as the positivism controversy, to criticize the critical rationalism of Karl Popper and Hans Albert.[89] For instance, Adorno's introduction to the collection of essays that documented the debate, *The Positivism Debate in German Sociology* (*Der Positivismusstreit in der deutschen Soziologie*, 1969), contrasts critical theory to scientistic approaches to sociology, which, according to Adorno, preserve an "innermost contradiction, unconscious of itself" in its combination of empiricism and logic.[90] Recall that, for Horkheimer and Adorno, the incongruence of empiricism and logic underpinned the return of logical positivism to metaphysics and the taboo of enlightenment that led it back to the violence of myth. After the war, mathematics still signified this contradiction: "Neither can one defend the absolute privilege of the individually given against 'ideas,'" Adorno writes, "nor can one maintain the absolute self-reliance of a purely ideal realm, namely the mathematical realm" (A 8.1:285–286). Note the persistence of terms: the "mathematical" marked the immanent failure of the logical positivists' approach to knowledge in the 1930s and pointed to the weakness in the scientistic approach to sociology in the 1960s. Those who relied exclusively on mathematics not only risked recapitulating the intellectual mistakes of the past but also were not critical theorists, in Horkheimer and Adorno's original sense of the term.

In the wake of the positivism controversy, Horkheimer and Adorno's equation of mathematics with logical positivism endured in the critical project, even as work by Habermas sought to reconcile critical theories of culture and society with the sciences. In contrast to Horkheimer and Adorno's equation of science with domination and control, Habermas's attempt, in the words of Axel Honneth, "to set the first generation of critical theory on methodologically solid ground" understood the sciences as working in the service of real and necessary human interests.[91] According to Habermas's *Knowledge and Human Interest* (*Erkenntnis und Interesse*, 1968), empirical-analytic sciences such as physics satisfy legitimate human needs, enabling us to predict and master the natural world through instrumental action. Habermas's point was to create a "critical philosophy of science" that escaped the illusion of the sciences' alleged objectivity and neutrality, upheld by the logical positivists and positivism in general.[92] And yet Habermas's more nuanced approach to scientific and technical instrumentality reiterates core elements of Horkheimer and Adorno's conflation of mathematics and positivism: as a theory of science, positivism "presupposes the validity of formal logic and mathematics."[93] To the limited extent we can even speak of it, Habermas's account of mathematics also emphasizes the exclusionary nature of positivism's exchange of mathematics for critical self-reflection on epistemology, which discounts other methodologies as "meaningless" and conceals "the problems of world constitution."[94] As it had for Horkheimer and Adorno, mathematics foreclosed for Habermas inquiry into not only the rational construction of the world but also the nature of rationality itself. While second- and third-generation critical theorists refined and expanded Horkheimer and Adorno's vision of critical theory, if mathematics remained in it at all it was as a symbol of an exclusionary and restrictive form of knowledge inseparable from the intellectual shortsightedness of positivism.

The critical image of mathematics set into motion by Horkheimer and Adorno's confrontation with logical positivism in the 1930s gave voice to a deep-seated yet persistent anxiety over the effects of mathematics on knowledge, culture, and art. To be sure: to reduce thought to mathematics is to exclude the qualities and materiality of people and things, to turn them into exchangeable and calculable units. Horkheimer and Adorno are thus correct that not all thought can be mechanized with mathematical symbols and operations. But, as the computer revolution has shown, some areas of thought can be manipulated—with powerful results—according to the rules of mathematics. On the one hand, Horkheimer and Adorno's antagonism toward mathematics limits critical theory's ability to respond to new forces of domination and control in a world ever more mediated by digital technologies—based, at least in part, on mathematics. On the other hand, it precludes the critical potential that mathematics offered some of Horkheimer and Adorno's intellectual precursors and friends—Gershom Scholem, Franz Rosenzweig, and Siegfried Kracauer. Writing mathematics out of critical theory may not have been the primary concern for Horkheimer and Adorno in the aftermath of World War II, as they reconstructed the Frankfurt School and worked to resurrect the legacy and gain recognition for the work of Walter Benjamin.[95] Nonetheless, the intellectual gulf opened up by their association of mathematics with instrumental reason, its role in the liquidation of philosophy and language, and, ultimately, the barbarism of the twentieth century has swallowed up recognition that theories of culture and art could draw on mathematics and remain critical at the same time. The same mathematics that Horkheimer, Adorno, and Benjamin criticized presented Scholem, Rosenzweig, and Kracauer with ways to critique and move past the intellectual impasses of their time. Their theories of aesthetics, messianism, and cultural critique, to which I now turn, bear the possibility of recovering mathematics' contribution to critical theory for the digital age.

The Philosophy of Mathematics: Privation and Representation in Gershom Scholem's Negative Aesthetics

For Gerhard (later Gershom) Scholem, mathematics unlocked the critical possibilities hidden in language. Mathematical logic, in particular, dispatched with the representational troubles introduced to philosophy through the everyday use of verbal and written language. "The foundational assumption that the ideas of concept, judgement, and the other basic elements of logic lie *beyond* phonetic language [*Lautsprache*], but within the sphere defined by the teaching of *signs* [*die Lehre von den* Zeichen]," Scholem wrote while studying mathematics at the University of Jena in 1917, "constitutes the legitimation of [mathematical] logic."[1] This new logic, in which words became mathematical signs and followed mathematical operations, promised to expand the philosophical horizons of traditional logic, perhaps paradoxically, by restricting language to "the teaching of *signs*." As discussed in chapter 1, it was such a reduction of thought to mathematics that, in the foundational texts of critical theory, not only threatened to liquidate language and philosophy but also drove Enlightenment's dialectical return to myth and barbarism. Accordingly, the elementary cognitive functions of calculation and equation that constitute the purest realm of mathematical thinking fulfilled Max Weber's dictum of modernity as the "demystification of the world" with disastrous results: things—both nature and other humans—can be dominated and controlled through mathematics.[2] For Scholem, however, the idea of stripping language of its usual representational features informed a metaphorics of structure and lack in his writing on the philosophy of mathematics that set the stage for what I call his theory of negative aesthetics. Scholem's negative aesthetics drew formal literary strategies out of the restriction of language in mathematics in order to turn language into a productive marker of its own restriction and to symbolize ideas and experiences that exceed these representational limits of language.

This chapter charts the emergence, development, and deployment of the metaphorics of structure lacking features of representation in Scholem's early work on mathematics, in which he received a university teaching degree (*Staatsexamen*) in 1922. This set of metaphors emerged from Scholem's interest specifically in the philosophy of mathematics, which studies the ontological and epistemological conditions that make mathematics possible. For him, the philosophy of mathematics represented the enduring possibility of knowledge in the midst of diaspora, in a time disrupted by war and rife with intellectual skepticism.[3] It constructed mathematical knowledge piece by logical piece, seemingly independent of the messy world of human affairs. In particular, the productive negative feature of mathematics resided in what Scholem came to see and describe as its privative (from the Latin participle *privatus*) linguistic structure which was able to represent but lacked the typical representational features of language such as phonetics and comparison; mathematics functioned, in Scholem's words, "without analogy" (*gleichnislos*, S 1:264). We have seen how such restriction underpinned the dialectic of enlightenment for Horkheimer and Adorno, and thus I turn here instead to the generative moment during World War I when Scholem's studies of mathematics and its elimination of representation in language entered his theorization of Jewish laments and his translation of the Book of Lamentations (*eikhah*) from the Hebrew Bible. How mathematics restricted the semantic function of language became, in Scholem's hands, the negative aesthetics of lament, which took silence, monotony, and rupture as a poetic strategy that gave voice on the level of form to the historical experience of privation and catastrophe.

In critical circles today, Scholem stands as the founder of the academic study of Jewish mysticism, but his early thoughts on the traditions and histories of the Jews were influenced by his studies of mathematics as well. The philosophy of mathematics helped Scholem move beyond an impasse in his thinking brought on by the destructive assimilative nature of the German-Jewish dialogue, often exemplified in his memoirs and diaries by his father.[4] The cost of the Enlightenment's emancipation of the Jews was assimilation, in which they renounced their cultural and religious traditions in order to participate in German society; a printer by trade and German nationalist, Scholem's father had readily paid this price, as Scholem recalls, working on Yom Kippur and lighting his cigar with the Sabbath candle.[5] His father's superficial observance of holidays and rituals threatened the accumulation and transmission of knowledge across generations of Jews, from the destruction of the Temples to the twentieth century. As with anarchist politics, mathematics offered a form of resistance—in particular, to his assimilationist father: "Despite it being Father's birthday yesterday (an event in itself), and all his children were present (predictably, the older he gets the more he wants to have his children around), I read the book by Voss that had just arrived, *On the Essence of Mathematics*, which is the book of a pure algebraist, meant for people without an eye for geometry [*das Buch eines reinen Algebraikers und für Geometrie ohne Organ seienden Menschen*]" (S 1:275).[6] The idea that many modern mathematicians foregrounded the syntax of algebra in exchange for the intuition of geometry also opened up the theoretical possibility that the apparent erasures of assimilation did not spell the end for Jewish traditions. Instead, by negating representation in language, "the essence of mathematics" revealed, at least for Scholem, the prospect that there could also be a Jewish tradition—the poetic genre of lament—that functioned not despite, but because of the contemporary negation of Jewishness.

By locating the possibility of tradition in its seeming negation, Scholem's work on negative mathematics responded to a particular crisis in the German-Jewish tradition, but it also opens up avenues in cultural criticism for thinking about the experience of exile and assimilation more broadly. Scholem's negative aesthetics ran parallel to the cultural work of his German-Jewish peers such as Franz Rosenzweig and Martin Buber, who sought to reestablish, in theory and praxis, a lost sense of Jewish identity through their translation of the Hebrew Bible (among other cultural ventures).[7] In contrast to Buber and Rosenzweig's efforts, which concentrated on restoring immediacy with the spoken word of God, Scholem's work on Jewish renaissance turned to the restriction of language repurposed from the philosophy of mathematics. Moreover, as Peter Fenves shows, concepts from the philosophy of mathematics addressed for Walter Benjamin a crisis in the inherited conceptions of language and translation, even as Benjamin eventually dismissed the critical efficacy of mathematics.[8] Scholem's aesthetics takes this underdeveloped contribution of mathematics' negative relationship to language to its critical conclusion. It shows that negativity—privation and apparent discontinuity—can become an epistemological and aesthetic ally of tradition, by transforming absence into a positive index, the erasure of expression as a symbol of deprivation. For example, in such works as *Major Trends in Jewish Mysticism* (1941), finding the continuity of tradition in apparent rupture became the paradoxical theory of history that Scholem tells through mysticism, a cultural practice that flourished at and was thus constituted by historical moments in which the Jewish people faced expulsion or destruction.[9] For traditions threatened by the homogenization of assimilation, as Scholem saw in Germany, this approach to negativity recommends strategies of mobilizing the limits of poetic language as a means of giving voice to experiences such as privation and loss. Scholem's work on mathematics thus helped shape the theological—even the emancipatory—dimension of restriction and refusal that persists in the critical project. For peoples living in exile, his aesthetics offers techniques for constructing a tradition out of silence and linguistic erasure, by turning writing itself into a symbol of deprivation and discontinuity.

This chapter takes Scholem's troublesome concept of tradition as its point of departure, where mathematical metaphors of structure and privation first emerge. For Scholem, tradition signified not only religions such as Judaism threatened by hyperrationality, secularization, and radical social politics but also the metaphysical paradigm governing the transmissibility (as in the

German, *Tradierbarkeit*) of knowledge as such. Tradition is, as Samuel Weber writes for Benjamin, the "structural possibility" of communicating thought between individuals and across generations.[10] As Scholem most poignantly formulates it in terms of mysticism in 1958: "The Kabbalist claims that there is a tradition of truth that is transmissible [*tradierbar*]. An ironic claim, because the truth under consideration here is anything but transmissible. It can be known, but not conveyed [*überliefert*] and precisely that in it, which becomes conveyable, is what it no longer contains. Real tradition remains hidden; only the decaying tradition decays on an object and becomes visible in its greatness."[11] This passage contains in nuce Scholem's critical contribution to the project of negative mathematics. Just because a tradition such as mysticism may appear to resist transmission does not imply its discontinuity; instead, such traditions may only be signified in the negative, as they "decay." The following analyses demonstrate how this metaphysical framework about mysticism developed for Scholem in dialogue with the philosophy of mathematics. Scholem's studies and early writings on mathematics set the stage for this dialogue as they circled around the metaphorics of structure, privation, and the restriction of language. The point of metaphorical transference came when Scholem's university studies in mathematical logic entered into his theorization of lament and translation of the five poems of the biblical Book of Lamentations, which recount the suffering of the Jewish people. The result, Scholem's negative aesthetics, is what mathematics' approach to negativity offers critical theories of history and tradition. Mathematics' venture to produce knowledge by limiting representation suggests to Scholem the continuity of discontinuity: the histories of diaspora and exile consist not only in moments of traditions' transmission but also in its seeming breaks, crises, and silences. This paradox constitutes the tradition of the Kabbalist, who can only transmit the immediate experience of the Godhead after the fact and through the imperfect medium of language, whose imperfection expresses the magnitude of the mystic experience. It also proposes a critical theory of tradition in diaspora, in which the experiences of rupture and catastrophe may seem, by all indications, to threaten the continuation of cultural traditions, while in truth the hidden core of tradition remains intact.

The Foundations of Mathematics and the Metaphorics of Structure

Scholem's negative poetics, his contribution to the project of negative mathematics, originated in his work on the philosophy of mathematics, which,
amidst the crises of assimilation and war, persisted for Scholem as a viable
source of knowledge. The philosophy of mathematics encompasses theoretical discussions regarding the nature of mathematical thinking, engaging
topics such as the foundations (*Grundlagen*) of mathematics as set theory
or formals systems and the relationship among mathematics, logic, and
language. Kant's *Critique of Pure Reason* set many of the terms of these
mathematical-philosophical debates, which, in the early twentieth century,
were picked up and continued by mathematicians and logicians such as
Gottlob Frege, Henri Poincaré, and David Hilbert.[12] A central philosophical question in mathematics that informed Scholem's negative poetics is
what Klaus Volkert calls the "crisis in intuition," rooted in the development
of new mathematical theories and fields in the nineteenth century, such as
non-Euclidean geometry, that evaded the usual reliance on spatial-temporal
intuition.[13] First, however, the philosophy of mathematics and proposed
solutions to such crisis translated in Scholem's early thought, which included
holding imaginary lectures on "the foundations of mathematics" (*Grundlagen der Mathematik*) on his way to and from the university, into a metaphorics
of structure and construction (as in the German term *aufbauen*). As war and
revolution raged in Europe, mathematics offered a stable epistemological
foundation on which knowledge, and the academic pursuits of a young intellectual, could build.

The foundationalism that the philosophy of mathematics provided for
Scholem drew on both disciplinary discourse and philosophical debates
reaching back to antiquity. Already in Plato, mathematics served as the pedagogical starting point for the higher education of philosophers, after their
elementary training in music and physical education. The ability to distin-
guish numbers and to calculate exemplifies the "common thing that all kinds
of art, thought, and knowledge use as a supplement."[14] Insights from mathematics also underpinned the pure forms of intuition in Kant's critiques,
which make synthetic judgments possible a priori, and form the key analogy
for pure thought for Hermann Cohen. Around the turn of the past century,

the idea that mathematics served as the basic building block of reasoning was tied to its increasingly central role in the strict formulation of the natural sciences, as in popular textbooks frequented by Rosenzweig: *Introduction to the Mathematical Treatment of the Natural Sciences (Einführung in die mathematische Behandlung der Naturwissenschaften*, 1901) by Walter Nernst and Arthur Schönflies (Benjamin's maternal great uncle).[15] As Voss, whose *On the Essence of Mathematics* Scholem snuck away from his father's birthday to read, explains: contemporary culture, in as much as it is concerned with the understanding and utilization of nature, finds its "actual foundation" (*eigentliche Grundlage*) in mathematics.[16] This position, that mathematics plays a foundational role in knowledge, constituted for Scholem a basis for his fledgling intellectual program, which he calls his "teachings" (*Lehre*) and "science" (*Wissenschaft*).[17] As he records in his journals at the beginning of his university studies: "I am still not sure whether the study of mathematics, to which I will devote myself, will make possible a starting point for my thinking from a mathematical standpoint [*eine Grundlegung meines Gedankenkreises vom mathematischen Standpunkt aus*], because the science is still temporarily closed off to me. For my best, I hope so" (S 1:177). This "hope" that mathematics could yield a "starting point" suggests that other academic fields could no longer provide such intellectual foundations. What mathematics affords is thus less a precise formulation of what thought or knowledge actually is than, in Scholem's words, a point of epistemological orientation for his theories of poetics and history.

For Scholem, mathematics supplied a starting point for thinking, because it exhibited unique epistemological and linguistic properties vis-à-vis other intellectual pursuits. He claims:

> Starting with Plato, all great thinkers have been mathematicians—consciously or unconsciously. Because, indeed, it is from here that something can be said: from mathematics and from history, where the last can only be taken in the highest and most complete sense, as if it were not, as in reality, subject to skepticism. Because these both are to be viewed as the path, as really the two foundational pillars of human spiritual life [*Grundpfeiler des geistigen Menschenlebens*] and the two singularly possible, eternal points of view, from which a starting point can be won: out of which one could essentially determine the concept of science. . . . Indeed, history is the unfree, mathematics the free

thought of science: because history fills one with disgust for the confusion that people call freedom, while mathematics fills one with the deepest joy of the necessary construction [*dem notwendigen Aufbau*]. (S 1:260)

Scholem's point here is that mathematics remains the only approach to knowledge that still produces findings that are beyond reproach and skepticism. In contrast to history, on his example, mathematics circumvents contemporary intellectual doubt because it operates "free" from the world of human history and according to its own inner logic. Mathematics thus produces knowledge not relationally, but structurally, providing—note the repeated emphasis on origins and structure—a "starting point" (*Ausgangspunkt*) and "foundational pillar" (*Grundpfeiler*) for thought. In this passage we see the emergence of a metaphorics of structure in Scholem's writing on mathematics, in that history and mathematics provide here the "pillars" that anchor thinking and that the latter offers through its form of a logical, "necessary construction."

If mathematics was potentially more epistemologically resilient than other subjects, then it could offer a framework to work through similar crises in modern thought. The skepticism that the passage above locates in history, often referred to as the "crisis of historicism," exemplifies a larger sense of cultural aporia in the late nineteenth and early twentieth centuries. As Kracauer writes in his 1921 essay "The Crisis of Science" ("Die Wissenschaftskrisis") this crisis of science consists in the belief that the statement that historical and social values are relative to the times and societies that produce them, as he diagnoses in the work of Ernst Troeltsch and Max Weber, is equivalent to a broad-based relativist outlook on the world, which is ultimately reducible to nihilism.[18] Such discussions of relativism, skepticism, and nihilism extended far beyond just the realms of history or science. They dovetailed on modernist suspicions surrounding the idea that language formed the basis for cognition and communication, which drew on Nietzsche's interrogation of language's relationship to truth and the language skepticism associated with the language crisis (*Sprachkrise*) of fin de siècle Vienna in the works of Fritz Mauthner and Hugo von Hofmannsthal.[19] In particular, Mauthner's *Contributions to a Critique of Language* (*Beiträge zur Kritik der Sprache*, 1901–1903), the reading of which inspired much of Scholem's early reflections on the relationship between mathematics and language, argues that language and thinking are at their core purely convention and, at best, possess an arbitrary relationship to reality or truth. As Benjamin Lazier explains, Scholem's early intellectual efforts concentrate on wresting a new type of synthesis from this sort of Nietzschean skepticism and nihilism, which for him is embodied in an "angel of uncertainty" that both haunted him as well as spurred him on to greatness (S 1:208).[20] It is significant that, in the face of intellectual crisis, Scholem retains the belief that knowledge is possible: "In my heart, I still believe in the possibility of knowledge [*an die Möglichkeit einer Erkenntnis*], I still believe despite all skepticism and all reservations—history, grammar, logic—on the justification of science" (S 1:138). As we shall see through the repetition of the singular form "a knowledge" (*eine Erkenntnis*), this synecdoche for knowledge hinges for Scholem on the starting point and foundation offered by mathematics.

While I will return to the relationship that Scholem's diaries posits between mathematics and history, it is first in the sphere of language and a theory of language that the initial kernel of knowledge afforded by mathematics begins to take shape. Indeed, the defense of language's epistemological efficacy is as much a counter-argument to Mauthner in the 1910s as it is a life-long interest of Scholem's.[21] The origin of this interest lies in part with Scholem's emergent interactions with Benjamin, in particular with Benjamin's 1916 essay "On Language as Such and on the Language of Man." The essay originated as an eighteen-page letter to Scholem that addresses such foundational themes as "mathematics and language, that is: mathematics and thinking, mathematics and Zion."[22] A well-spring for contemporary scholarship and theories of language and criticism, "On Language as Such" argues in the main against the instrumental and the mystical conception of language—the former, "bourgeois" conception reducing language simply to an instrument of communication and the latter conflating language with the mystical experience itself. As an alternative, and in conjunction with Kant and Cantor's writings on mathematics, the essay postulates three distinct and infinite "orders" of language: the language of God, humans, and things.[23] These three languages interconnect, in that humans and things are created by God's word and human language participates in this divine language by giving names to things. The consequence of Benjamin's conception of language is language's infinitude. Language as such thus has infinite permutations, meaning that "human" language makes up only a portion of language as a whole and what the infinitude of language can communicate.

The key conceptual ingredient that Benjamin's essay offered can be found in its emphasis on the name, which revealed to Scholem the epistemological efficacy of mathematics to break through the impasses posed by skepticism and crises in tradition. In a certain sense, Benjamin's conception of language agrees with Mauthner's: a fundamental incongruence separates an object from the language we use to describe it, which "On Language as Such" identifies as the fall of language. And yet, for Benjamin, it was not always so. According to the essay's interpretation of Genesis, God brings things into being through the creative word so that they are recognizable (*erkennbar*), and we give them names in that we recognize or "know" them (as in *erkennen*). Hence, the act of giving names (*benennen*) to things forms the linguistic essence of humans, and it is in the original, Adamistic form of the name in which objects and words coincide.[24] Naming thus differs from the arbitrary designation of meaning (*bezeichnen*) that devalues mathematics in Benjamin's eyes (see chapter 1). What is unique (indeed, magical on some accounts) about Benjamin's idea of naming is how it allowed a type of intentional and self-reflexive signification that represents simultaneously itself and its putative object. One can think of the name as a semantic onomatopoeia, as a sign that mimics and discloses not its sound, but its own mechanism of signification.[25] Although the act of naming (*benennen*) allows recognition (*erkennen*) for Benjamin in language, the creative epistemological force of mathematics resides for Scholem in mathematics absolute lack of "names": "A highly significant remark, which follows necessarily from my way of seeing things and which immanently operates *very* strong in Benjamin: *definition is knowledge* [*Definition ist eine Erkenntnis, ist die Erkenntnis*]. Everything else is interpretation of the definitions. Mathematics is the nameless teaching [*die namenlose Lehre*]: is knowledge, indeed metaphysical knowledge" (S 1:467). Given the repetition of the singular synecdoche for "knowledge" (*eine Erkenntnis*), the passage suggests that, against Maunther, the definition fulfills the same epistemic function for Scholem that the name does for Benjamin: it provides a foundation for knowledge. In contrast to Benjamin's concept of the name, however, the epistemic power of the definition hinges on arbitrary signification, that is, on its "metaphysical" independence from the object it signifies—to which I turn in the next section. Note also how the passage already reinforces the sense of mathematics as a structure of definitions and interpretations, characterized, in particular, by privation ("name-less").

This metaphorics of structure in the philosophy of mathematics were the initial theoretical steps that, in the course of a year, lead Scholem to a negative aesthetics, but not because mathematics provided a rigorous justification for his emerging "circle of thoughts." Instead, what Scholem's work with the foundations of mathematics revealed was the simple structural possibility of knowing, the potential continuation of the Enlightenment project. Despite a crisis in tradition and the popularity and cultural weight of skepticism and nihilism in science, language, and history, there remained the possibility for knowledge (*Erkenntnis*) in mathematics that did more than just reiterate the limits of language. This possibility emerged in metaphors of structure itself—terms that, by no coincidence, bear a resemblance to those used by Carnap and other logical positivists.[26] For Scholem, then, the possibility of knowledge lay less in Benjamin's concept of the name in which subject and object intersect than in a theory of the arbitrary, mathematical definition, which eschewed language's usual function of representation. At Scholem's admission, the idea that a simple mathematical definition constitutes knowledge may seem trivial and, indeed, what Scholem means by the definition still lacks a certain definitiveness (S 1:467). This seeming lack of value or clarity, however, does not detract from what mathematics offers in terms of knowledge—instead, this sense of lack derives from the very nature of mathematics' contribution to epistemology.

"A Great Tautology": Negativity in the Privative Structure of Mathematics

If the philosophy of mathematics guaranteed the possibility of knowledge through a metaphorics of structure, then its core structural characteristic consisted for Scholem in privation, its lack of connection to nonmathematical thought. Indeed, this troublesome relationship of mathematics to the rest of the world presented a mathematician like Poincaré, one of Scholem's key interlocutors, with an "irresolvable contraction" of mathematical reasoning. One the one hand, according to Poincaré, mathematics purports to be a logically deductive science. Its source lies in the consistent basis of logical inference aided only by arbitrary signs, which affords mathematics the widely acknowledged epistemological status of "complete irrefutability."[27] On the other hand, this sort of deductive, logical reasoning detached from experience struggles to come up with anything "essentially new," beyond that which reduces to the identity principle, "A equals A." All of mathematics thus equates to one massive tautology. We have seen this position, that mathematics is an assembly of self-referential logical statements, held by the logical positivists, which amounted, as Horkheimer and Adorno objected, to the reduction of thought to immediacy and repetition. In Scholem's writing on the philosophy of mathematics, however, the idea of mathematics as a giant tautology shaped the metaphorics of privation. In our exploration of negative mathematics, this structure—defined by lack, independence, and inaccessibility—constituted the generative negative element of mathematics for Scholem. It revealed mathematics' ability to create knowledge despite what seems like privation from a human perspective, opening up a metaphysical framework in which poetics and tradition could potentially generate knowledge as well, in spite of the deprivation of exile and the erasure of assimilation.

The question of whether mathematical judgments borrow information from the nonmathematical world or are all just self-referential tautologies received its decisive formulation in Kant's critical philosophy. His *Critique of Pure Reason* set many of the terms of the debate over the nature of mathematical thinking at play throughout this book. To briefly review his argument from the "Transcendental Aesthetic": for Kant, judgments that draw from experience are a posteriori and those that do not—that is, pure and transcendental judgments—are a priori; likewise, he calls judgments that unpack what is already in a concept *analytic*, whereas those that integrate new information are *synthetic*. Countering the skepticism of David Hume, the first *Critique* seeks to delineate the possibility of synthetic judgments a priori, which lies between Hume's designations of the objects of reason as "Relations of Ideas" (analytic judgments a priori) and "Matters of Fact" (synthetic judgments a posteriori).[28] The prime example of this special category of judgments—which integrate new pieces of information, but which are "pure" in that they precede the empirical—can be found, Kant believes, in mathematics in general and in algebra and geometry in particular.[29] For Kant, mathematical judgments are clearly a priori; they "carry necessity with them that cannot be derived from experience."[30] And yet a judgment in algebra ("$7 + 5 = 12$") or in geometry ("the straight line between two points is the shortest") is also synthetic, because it relies on an extra source of information, the idea of equivalence or that of shortness taken from time and space. This argument followed from Kant's self-proclaimed Copernican Revolution: space and time are not products of empirical experience, but rather pure forms of intuition, the cognitive conditions that make experience possible. Mathematicians disputed Kant's claim of the synthetic nature of their subject and Neo-Kantians later rejected space and time as pure forms of intuition.[31] This post-Kantian debate over the synthetic or analytic nature of mathematical judgments provided the context in which Scholem's ideas about the autonomy of mathematics and its privative structure emerged.

The self-containment of mathematics stems for Scholem from the belief that mathematical judgments do not borrow any information from nonmathematical sources, the so-called pure forms of intuition included. Scholem's conviction of the independence of mathematics is perhaps most evident in his forceful disputation of the term *philosophy of mathematics* itself (even if this is how historians of mathematics later classified Scholem's areas of mathematical interest).[32] The foundations of mathematics and the essence of mathematical reasoning cannot, according to Scholem, be expressed in the foreign "jargon" of philosophy, because they constitute two separate modes of cognition (S 1:259). The creation of his "new science" (*neue Wissenschaft*) to overcome the nihilism and skepticism of Nietzsche and Mauthner hinges not on the "philosophy of mathematics," but rather on the "mathematics of mathematics" (S 1:258–259 and 264). What is striking about this distinction is not only its reference to the German romantic poet-philosopher Novalis but also the self-reference it ascribes to mathematical reasoning.[33] This notion expands and defines the metaphorics of structure in which Scholem writes about mathematics: the structure of mathematics and the knowledge it creates exist and interact in a realm of their own, restricted, and only partially accessible to the nonmathematical world, including philosophy.

As an example of Scholem's insistence on the independence of mathematics, consider his objection to Poincaré's theory of mathematical induction. For Poincaré, induction was the only possible synthetic a priori judgment.[34] The core idea of induction is that algebraic statements such as $2 + 2 = 4$ consist of the recurring operation of $x + 1$, executed two, four, or an arbitrary number of times. Given the lawfulness of human understanding, such a result can be generalized into the claim that, given a statement, if the case n of the statement being true implies that the case $n + 1$ of the statement is also true, then the statement is true for the cases of all natural numbers $(1, 2, 3, \ldots)$.[35] For example, if I can prove that the sun rose today and if I can prove that the sun rising any day in the future (day n) implies that it will also rise the next day (day $n + 1$), then I have proved that the sun will rise every day, from today onward. New information thus enters, according to Poincaré, the otherwise tautological structure of mathematics. For Scholem, however, induction imports "an aid taken from a foreign domain, not belonging to mathematics," relying on the idea from philosophy and psychology of a "potentially infinite imagination" (S 1:268). Such methodological borrowing from philosophy sold short the unique epistemological contribution of mathematical reasoning for Scholem, threatening the purity that made mathematics a structure of "free" construction immune to the skepticism of subjects like history. This idea is key: if the generative aspect of mathematics lies in its lack of relationship to the nonmathematical world, then perhaps there are languages in which aesthetic and cultural traditions can produce and transmit knowledge even when their relationship to the outside world has become problematic.

In particular, the epistemological contribution made by mathematics resides for Scholem in this negative relationship to the nonmathematical world: mathematical knowledge arises not in relation to experience, but instead through the absence of relation altogether. Clarifying this lack of relationship helped Scholem come to a resolution of the synthetic-analytic debate; it also underscores the creative element of negativity at work for Scholem in mathematics:

> The expression, that mathematics is a great tautology A = A, has really nothing off-putting about it, if only understood correctly: mathematical propositions

and truths have all already been there since eternity, infinite mathematics is, as paradoxical as it sounds, indeed completed. Of concern is only that the human mind [*menschlicher Geist*] knows every single one of these infinitely many propositions through the logical connection of the already known. Maybe the propositions really express nothing but what lies in the definitions, but in these [definitions] lie an entire world folded together. The assignment of mathematical thinking is just to unfold them.

He elucidates this point with a comparison illuminating its proximity to mysticism:

All the wisdom of the world lies folded together in the twenty-five letters [*sic*] of the German language and it requires only the—admittedly "creative"— combination to derive *Don Quixote* out of it, which, as seen from here, is an analytic piece of wisdom. And in the fact that there are infinitely many combinations, as is easily understood, lies precisely that one can-*not* designate mathematics as tautology in human language [*Menschensprache*] (from God's perspective, sure!), because precisely the wealth of truths can *never* be exhausted. Thus, mathematical knowledge can never come to an end; thus, ever anew will already existing truths be found in the "eternal empire of ideas." *An infinite tautology is, as seen by humans, not a tautology*—that is the crucial point. (S 1:277–278)[36]

Two main ideas are at work in these passages. The first is the claim that, for Scholem, mathematics, "a great tautology," consists entirely of analytic judgments, "A=A." Both passages thus position him in line with the mathematical-philosophical perspective that views mathematics as an analytic construction unfolding from initial "definitions" as, in its different formulations, logical statements (Frege), formal axioms expanded by logical inferences (Hilbert), or the combinations of a universal alphabet of human thought (Leibniz).[37] The second idea is mathematics' structure of privation that distinguishes metaphysically between "God's perspective" and "the human mind." For Scholem, "the divine mathematician," as he puts it elsewhere, comprehends the "infinite" totality of mathematics, while, "in human language," the assignment of "mathematical thinking" lacks conclusion: "the wealth of truths can *never* be exhausted" and "mathematical knowledge can never come to an end."[38] This is the negative element of mathematics that held critical potential for Scholem: humans will never exhaust the infinity of mathematical truths, but can, nonetheless, gain mathematical knowledge by eternally unfolding its "infinite tautology."

Scholem's reference to "definitions" and "human language" in these two passages is further illuminating, because it suggests that the relationship between mathematics and language produces knowledge in mathematical thinking despite its privative structure. Around 1900, however, the definition was a disputed philosophical concept. For instance, the definition epitomizes the epistemological poverty of logic according to Mauthner: it takes part in a "societal game" (*Gesellschaftsspiel*) that either depends on the point of view of the subject (*vom Gesichtspunkt abhängt*) or unfolds as a "tautological examination" of a concept we all already know.[39] In contrast, as Scholem hinted in his response to Benjamin, human mathematical thought creates knowledge of the infinite world of mathematics through the definition:

> I cannot go along with Mauthner's critique of the definition, that definitions are always tautologies. Because of mathematical definitions. The definition of a [straight] line is not a tautology, because a word which is entirely without meaning: LINE is rendered meaningful in connection with certain intuitions [*Anschauungen*], where one could just as easily (come up with) a different definition that would result in a completely different concept. Mauthner would have to claim that the definition only expresses what we somehow already know about a line. Sure, but how and from where do we then know something about a line, which after all is a fictitious, unreal ideal concept [*ein erdichteter, unwirklicher Idealbegriff*]. The definition is indeed meaningful here, even if it is not a synthetic judgment, because it adds nothing to the concept "line" that was not already in it, but rather only and merely expresses that which should be understood under the concept. (S 1:139)

The passage insinuates not only a division but also a linkage point between mathematics and Scholem's developing ideas regarding language. The division follows the independent, metaphysical nature of mathematics: while we can associate its objects with a given "intuition" (*Anschauung*), they are ultimately, as with the "straight line," ideals beyond reality (*unwirklich*), fictions invented (*erdichtet*) for, and independent of their use in human language. For Scholem, then, the definition served as the liminal point between this ideal world of mathematics and language; it provided knowledge about a "straight line" by giving linguistic expression (*sagt . . . aus*) to the pure mathematical idea.

The "definition," in the way it supposedly functioned for Scholem, gains more definite contours in comparison to the properties Kant ascribes to the definition in mathematics and in contrast to Benjamin's concept of the name. For Kant, the core difference between philosophical and mathematical reasoning is that the former represents cognition that follows rationally from concepts, while the latter, drawing on the pure forms of intuition, first must synthesize or construct its concepts.[40] Hence, if to define something means "to exhibit originally the exhaustive concept of a thing within its boundaries," then a definition in philosophy is an "exposition of given concepts."[41] According to Kant, something different takes place in mathematics: "mathematical definitions can never err. For since the concept is first given through the definition, it contains just that which the definition would think through it."[42] Scholem's concept of the definition picks up on and transforms Kant's, as is clear in the previously cited objection to Mauthner. For Scholem, the definition is itself not a synthetic judgment a priori. Instead, Scholem writes, "[the definition is] an arbitrary naming. The thing, which is only there once between two points, we name—whether it exists or not—we call it straight line as a start."[43] Scholem's standpoint differs from Kant's in where they locate the creative cognitive moment: for Kant, it comes when pure intuition "gives" us the objects we cognize; for Scholem, it is in the linguistic act of defining that "expresses" and "names" (*nennen*) its object. Furthermore, we see again the similarity and difference between Scholem and Benjamin: both locate an origin for knowledge in language, but for Benjamin, this act (naming) was meaningful and necessary, whereas for Scholem, defining was stipulative and arbitrary. The specific arbitrariness of the mathematical definition was significant for Scholem, not only because it signaled mathematics' special relationship to language but also because it suggested that language carries with it a symbol of its own limitation.

Scholem's vision of mathematics as a "great tautology" embodies critical potential beyond the realm of mathematics, establishing a connection in the negative between "human language" and knowledge that lies beyond it. The notion that mathematics arises mechanically and analytically out of the statement "A = A" also serves as a theoretical dividing line, at least in the area of mathematics, between Scholem and not only Benjamin but also Horkheimer and Adorno. In the context of Benjamin's "On Language as Such," Scholem called mathematics "the nameless teaching" (S 2:213); the critical project envisioned by Benjamin in *The Origin of the German Tragic Drama* opposed this lack of names in mathematics as an abandonment of representation and meaning (see chapter 1). Here Benjamin as well as Horkheimer and Adorno missed the positive negativity that mathematics offered Scholem, the latter two instead equating the restrictive features of mathematics with neurotic regression and the ritualistic repetition of myths. For Scholem, however, mathematics was not the instrumental application of number to thought and nature. Instead, mathematics delineated its own form of representation and field of knowledge that provided insight into how representation and knowledge work when presented with privation. In the face of this giant tautology, mathematicians found alternative means in order to represent a field that, for Scholem, only God can know in its totality. How mathematics contorts language and representation suggested to Scholem ways in which aesthetics and cultural traditions can also employ such contortions to express privation and lack.

Mathematical Platonism and the Limits of Language

For Scholem, investigating the relationship between mathematics and language completed the metaphorics of structure and lack in the philosophy of mathematics by specifying the element that was absent in mathematics: representation. Mathematical-philosophical debates over the ambiguities of language provided the context. Bertrand Russell and Alfred North Whitehead, for instance, developed a symbolism for mathematical thinking to avoid the imprecision of "ordinary language"; Gottlob Frege, with whom Scholem briefly studied, developed a "concept notation" (*Begriffsschrift*) to overcome the "inadequacies of language" and avoid the ambiguities of representation.[44] Accordingly, arbitrary symbols and lines designating relation communicate more clearly and concisely mathematical knowledge than the languages of English or German. For Scholem, the move from language to arbitrary symbols met the Platonic world of so-called "mathematics as such" half way, not adding anything through language to "this pure mathematics" that is "absolutely logical, analytic, and here since eternity" (S 1:427). This move indicated to Scholem the generative aspect of negativity in mathematics, coded in a metaphorics of structure lacking the figurative components of language, such as analogy. The move also matched mathematics to another of Scholem's interests, mysticism. Mathematics and mysticism came to form two sides of the same coin, both expressing what lies beyond human experience; but where mysticism captures this knowledge in language, mathematics expresses it by restricting language. This is the point at which Scholem's thinking on the philosophy of mathematics and his theorizations of aesthetics and culture start to collide.

The purity of mathematics referred, according to Scholem, not only to its independence from the nonmathematical world but also to its eschewal of representation in language. This notion of purity picked up and expanded on the Kantian tradition. For Kant, *pure* served as the key term in his discussion of space and time, which are not simply things I experience empirically, but rather are the "pure" cognitive forms that render my empirical experience possible. For a Neo-Kantian such as Cohen, the task of "pure" thinking in his 1902 *Logic of Pure Knowledge* was to eliminate from thought what the senses deliver to us as untrustworthy perceptions of the empirical world.[45] According to Scholem, mathematics is "pure" in both terms of experience and language:

> As mentioned above, mathematics distinguishes itself from all of the many other pursuits that one erroneously counts as science primarily through one thing: through its lack of analogy [*Gleichnislosigkeit*]. There are several things to say about this: it is nearly self-evident in language to speak in analogy, in symbol: most often that what is known cannot be said at all other than in symbols. Open any book: one finds everywhere the formulas of analogic speech: as if, just as, and similar expressions. The core cannot be said: nature, because nature is unsayable, rather it can only be alluded to imagistically, the pseudo-science is essentially allusion to an inexpressible fact [*eines unaussprechlichen Tatbestandes*], which can be experienced by humans and therefore is solely accessible through the medium of speech in analogy. Entirely different here is mathematics, which wants something entirely different, unheard-of, which puts mathematics in relation—in its goal—and in sharpest contrast—in its

path—to the second great possibility: mysticism, which sees its unscientificity as its essence.[46] (S 1:264)

I will return to the link between mathematics and mysticism, but what is striking in this passage is how it depicts mathematics through a privative judgment, in the Aristotelian sense of the term. In contrast to a judgment of negation ("A is not B"), a judgment of privation asserts that "A" lacks an attribute "A" normally possesses: while many branches of knowledge and science usually use analogies, mathematics is the "science" and "activity" defined by its lack (as with the suffix *-losigkeit*) of analogy (*Gleichnis*). According to the passage, lack (*-lessness*) is not a source of epistemological impotence, but rather the attribute that allows mathematics to produce knowledge (grasp a "core" or "nature") where other modes of knowledge resort back to the use of symbols, comparisons, and allusions.[47] The passage even formalizes lack in that it leaves analogy ironically undefined and adrift among such other terms as *symbol, formula,* and *allusion,* suggesting that mathematics' advantage lies in its capacity to function without the confusion that language introduces through representation.

Associating mathematics with such privation is not an arbitrary choice on Scholem's part, but instead positioned Scholem in contemporary philosophical debates in mathematics, if not epistemology as a whole. In his musings on the subject, Scholem even "has the vague premonition" that he is heading in the direction of mathematical "Platonism" (S 1:278).[48] Mathematical Platonists believe that mathematical objects (such as numbers, functions, or sets) exist and that their existence is independent of the human mind and language.[49] In contrast to mathematicians such as Richard Dedekind and Aurel Voss, who believed that a mathematical object such as a number was a free creation of the human mind, Platonists would contend, to cite Scholem's example, the idea of number exists independently of human modes of cognition and representation.[50] Although humans may play a role in "inventing" numbers (*erfinden*), we merely "discover" (*entdecken*) them, just as Columbus "discovered" an America that already existed well before the arrival of Europeans (428). As humans, mathematicians simply give arbitrary signs to mathematical concepts. As a whole, according to Kurt Gödel, mathematics describes "a non-sensual reality, which exists independently both of the acts and [of] the dispositions of the human mind and is only perceived, and probably perceived very incompletely, by the human mind."[51] As Scholem's definition emphasizes lack (*-lessness*), the Platonist conception of mathematics also builds on words that hinge on lack and privation, such as "very incompletely," "independently," and "only." To be clear, this Platonist "non-sensual reality" of mathematics is neither the Messianic Kingdom nor the mystical experience of the Godhead. Instead, mathematics and mysticism both attempt to express "an inexpressible fact"—mathematical reality and God, respectively—that exists in and unto a world independent of the very strategies of representation at our disposal to describe it.

More than a superficial commonality, the affinity between mathematics and mysticism articulated for Scholem the negative element in mathematics that serves as an epistemological ally to push through crises in tradition and history. He saw, however, a significant difference between the two in the degree to which they rely on the mechanism of representation in language: mysticism is the complete saturation of representation and mathematics its total absence. As his diaries explain:

> Mathematics and mysticism: the core of both stands the test through the following: it is attempted, or much more, sensed as a self-evident assignment: to express the unity of the world, to express it in its essence. For that such a unity exists is, as a "philosophical" axiom, the foundation of everything. And precisely here the great antithesis reveals itself: mathematics can speak only naked, without analogy, mysticism only in image and analogy. For mysticism takes up a unity in its totality that is inaccessible to all knowing language, but mathematics rebuilds a broken-up but perceived unity in its own way. (S 1:265)

In contrast to mysticism, mathematics lacks for Scholem not only the comparative element of language ("analogy") but also its rhetorical rules and strategies; it speaks "naked" and free of "images." The passage thus affords mathematics a special status as representation that "expresses" knowledge by restricting the normative features of language. As the metaphorics of structure ("rebuilds") lacking analogy suggests, mathematics constitutes, to borrow a term from Hans Blumenberg, "absolute metaphors": like God and truth, mathematics serves for humans as an irreducible "translation" of an object—"the unity of the world," Platonic "mathematics as such"—to which no perception corresponds.[52] Analogy and figurative speech are absent in mathematics for Scholem, because they presuppose an idea of the object that they illustrate or represent. In contrast, the generative negativity of mathematics lies, according to Scholem, in its absoluteness as metaphor: mathematics produces and transmits knowledge in absentia of its objects, as we previously saw, through their arbitrary definition and logical interpretation. Mathematics' privative structure even resembles on the level of negativity the Platonic world it describes, in particular, its ultimate isolation from human thought and language.

Scholem's discussions of how his contemporaries related mathematics to philosophy and mysticism offer two examples that help clarify the epistemological contribution made by mathematics as absolute metaphor. This nuance will help later differentiate Scholem's employment of negativity in mathematics and mathematical thinking from the other contributions to negative mathematics explored in this book. First, consider the difference between Scholem's conception of mathematics and Cohen's, which Scholem deems a "foolhardy perspective" (S 1:261). In Cohen's *Logic of Pure Knowledge*, mathematics illustrates via analogy the possibility of pure thought: infinitesimal calculus constructs its objects without recourse to intuition or empirical givens. To cite an example from Cohen that Scholem finds particularly problematic: "The coordinate axes form an important representative [*Vertretung*] of the thought of substance" (276).[53] Chapter 3 shows that concepts from mathematics such as infinitesimal calculus thus illustrate for Cohen pure thought, not because they are the mathematical avant-garde, but rather because by the end of the nineteenth century, mathematical ideas such as infinitesimal calculus were widely accepted and understood.[54] Although we understand the idea of pure thought, mathematics provides, in Cohen's and, later, Rosenzweig's work, an analogy for the inner-working of pure thought, which, for nonphilosophers, may be more difficult to understand. In contrast, mathematics eschews for Scholem rhetorical strategies like analogy, because language not only has no bearing on the Platonic world of mathematics but also only serves to obscure it.

The second example concerns Scholem's objections to the use of mathematics by mystics to represent and stand in as a comparison for the otherwise incommunicable. Scholem was positioning himself here against mystical links between mathematics and occult forms of knowledge, exemplified for him in the writings of mystical thinkers, including Novalis, Oskar Goldberg, Martin Buber, and Rudolf Steiner.[55] For instance, Steiner maintains in his speech "Occultism and Mathematics" ("Okkultismus und Mathematik," 1904), empirical mathematical objects (*Gebilde*) only refer, serve as the "analogy" (*Gleichnis*) in experience for a "spiritual fact" (*geistige Tatsache*). Through training in the spirit of the mathematical, occultists could find one path toward cleansing themselves of the life of sensuality.[56] Scholem would find Steiner's employment of mathematics in the service of attaining occult-mystical knowledge problematically superficial, because according to it, mathematical objects naively step in for the meta-sensual strived for in anthroposophy. As such, numbers, in Scholem's words, would be the letters of Galileo's book of nature and, as a whole, mathematics would provide us access to the incommunicable as such (S 1:407). But Scholem was a student of mathematics and, later, a historian of mysticism. Although he often played with the mathematical-mystical calculations of Gematria, this line of thinking entailed and employed mathematics as either a mimetic corollary to or an analogy for the secret knowledge purported by mysticism—both of which ran against Scholem's stricter conceptions of mathematical thought.[57] In contrast, mathematics and mysticism were interrelated, not because mathematics offered the secret language of the incommunicable postulated by mysticism. Rather, they were interrelated because they both spoke to the general and more salient logical perplexity (as shared "in their goal" in the passage previously cited) of representing that which exists independently of representation. The difference was that mysticism spoke to this perplexity by proliferating signs, while mathematics spoke to it—and this is the key to the next section—by restricting signification.

There is a striking similarity between Scholem's position that mathematics is a structure lacking representation and his concept of tradition. Recall from the passage cited at the start of this chapter that the tradition of the Kabbalist consists, in his words, of a "real" core and its "decaying" instantiation in language. In Scholem's framework, mathematics likewise consists of arbitrary definitions followed by logical construction, with the "real" core independent of thought and representation and knowable in full, in this view, only to God.[58] If mathematicians create mathematical knowledge by defining and interpreting the Platonic world of mathematics, then tradition can function according to the same logic: a tradition, take Judaism, is defined in the Torah as the absolute word of God, which each generation accepts, interprets, and passes on. This is not to say that mathematics illustrated or was itself God's word—in fact, Scholem argues convincingly to the contrary (S 1:468). Instead, mathematics and a tradition like mysticism both ventured to describe realities that for Scholem exist beyond the limits of human mind and language: the mathematician and the mystic produce and transmit knowledge of their subject matter, even if this exists only in a "decaying" state. Indeed, negativity in mathematics—located for Scholem in the metaphorics of structure lacking representation—suggests that this "decaying tradition" also serves as a marker of the fact that "real tradition remains hidden" beyond the limits of mind and language. The epistemologically generative eschewal of analogy in mathematics—its negativity— thus affirmed the possibility that some traditions may function not despite, but because of privation.

From Mathematical Logic to Jewish Lament

The year after Scholem read about the essence of mathematics on his father's birthday, he held an in-class presentation (*Referat*) at the University of Jena that served as a key transition point between his works in mathematics and his theorization of aesthetics and tradition. Presented in Bruno Bauch's seminar on logic, Scholem's *Referat* defended mathematical logic ("Logistik"), the translation of logic into mathematical symbols and operations, against its detractors in philosophy, namely Hermann Lotze's *Logik* (first published in 1843).[59] In the weeks following the *Referat*, Scholem refocused his creative energies on another intellectual passion, Judaism, by translating the Book of Lamentations (איכה or *Klagelieder*) from the Hebrew Bible. Here we see Scholem's prime contribution to negative mathematics: through the metaphorics of structure defined by the restriction of representation, these theorization of lament as a poetic genre and translations of the biblical lamentations into German transformed the philosophy of mathematics' approach to negativity into a creative literary strategy. Keep in mind that this same disavowal of representation indicates, for Benjamin and the Frankfurt School, how mathematics excluded linguistic mediation, leading back to myth. In Scholem's work on lament, however, restricting representation became a way of representing in negative, through a formal kinship of semantic absence in mathematics and poetic language. In the aesthetic forms of silence and monotony, the lack of representation paradigmatic in mathematics could signify the hardship and deprivation of the Jewish people that Scholem's translations lament. For critical theory, Scholem's negative mathematics offers literary strategies that do not represent the unrepresentable, but rather indicate that the loss of diasporic peoples and the erasure of tradition through assimilation often exceed the limits of language.

Scholem's study of mathematical logic built the bridge between his work on the philosophy of mathematics and lament. Mathematical logic attempts to clarify the problematic but also highly productive relationship between mathematics and logic. Around the middle of the nineteenth century, mathematics and logic formed two distinct branches of knowledge, yet by the 1850s logicians such as George Boole undertook measures to push logic past the traditional limits of Aristotelian logic—a development resisted by some philosophers and logicians such as Lotze.[60] To expand logic past syllogistic reasoning, Boole's algebra of logic translates logical statements into suitable symbolic-algebraic equations, manipulates these equations with the help of algebraic operations, and translates the results back to the language of logic. Take, for instance, the example Scholem would have encountered in Lotze: "the fundamental law of thought" is for Boole represented by the equation $x^2 = x$, which, through a few simple algebraic operations, equals $x(1-x)=0$, the principle of noncontradiction. Instead of drawing conclusions based on the linguistic statement, in Boole's words, "it is impossible for any being to possess a quality, and at the same time not to possess it," we write in symbols $x(1-x)=0$, which can be manipulated to derive further results using the rules of not language, but rather algebra.[61] Mathematicians and logicians such as Frege, Russell and Whitehead, and Giuseppe Peano developed symbolic notations similar to Boole's; together, these systems of logic are referred to as mathematical logic, logical calculus, and "Logistik."[62] Likewise, mathematical logic expresses mathematics through logically grounded axioms, logical rules of inference, and, above all, a neutral and formalized language of symbols. To be sure, any attempt to eliminate language appeared intellectually dubious to Scholem; his *Referat* hastened to emphasize the limits of mathematical logic, which forfeited questions of history and religion (S 2:66 and 111). But mathematical logic revealed to Scholem that there were other ways to communicate that not only work beyond the usual rhetorical and semantic structures of language but also employ the absence of rhetoric and semantics to their epistemological advantage.

The *Referat* provided a counterargument to Lotze's critical assessment of attempts to formalize knowledge into arithmetic-mathematical syntax and logical operations, from number mysticism to Boole to more recent mathematical-philosophical trends in the German academy. While the bulk of the *Referat* addressed Lotze's discussion of the careful—and, for Scholem, excruciatingly longwinded—application of the a priori principles of thought in language, it primarily disputes Lotze's claim that Boole's mathematical logic simply tells us something we already know (an objection analogous to Mauthner's criticism of the definition).[63] Where Lotze saw the unnecessary repetition of logical statements in mathematical syntax, the *Referat* finds the possibility of an inroad into how a limited view of language can be generative. It explains:

> A real contestation of "Logistik" could only be based on evidence that logic has a language, which, on the one hand, is most intimately connected to phonetic language [*Lautsprache*], but, on the other hand, would be representable not without remainder [*restlos*] in written signs. Yet there is no prospect whatsoever, that such a proof can ever be delivered. In fact, the idea on which in the end the entire edifice of mathematical logic rests seems to have a lot going for it: *that pure thought* [reines Denken] *can only be represented without remainder in pure symbols.* (S 2:110)

On the surface, the passage states Scholem's support for Frege and Russell and Whitehead in, respectively, *Begriffsschrift* and *Principia Mathematica*, which in the case of the former strives in fact to create a "formula language" (*Formelsprache*) of "pure thought." But the passage also undertakes the *Referat*'s first philosophical step in that it answers the question what a "pure symbol" may be. Again, the term *pure* here deviates from the Kantian connotation: independent of experience. Instead, and via a markedly and, at

points, confusingly negative vocabulary ("no prospect" and "not without remainder"), the concept of purity builds on the metaphorics of privation,
like mathematics as a whole. The pure symbolism of the pure thought of
"Logistik" assumes that there are no elements of thought that cannot be expressed in pure symbols (that its success depends on the lack of a "remainder"). Furthermore, these symbols can be fully decoupled from "phonetic
language," language enunciated out loud and language based on phonemes
as the smallest units of meaning in speech. According to Scholem, the success or failure of "Logistik" thus lies in its ability to realize the constriction
of representation—written and spoken—that he posited as the essence of
mathematics above.

The *Referat*'s next philosophical move addressed the question that Poincaré raised at the beginning of the previous section: how mathematical logic
can become "fruitful" beyond the "pure" tautological structure of mathematical reasoning.[64] Scholem's answer drew on and expanded Benjamin's
idea of language's infinitude, reaching conclusions that must have alienated
his listeners.[65] The *Referat*'s solution to the problem of how mathematical
logic may be fully realized, hinges on the idea that there may be other forms
of language beyond human language:

> The often-raised objection [against "Logistik"], that the principles and
> ur-symbols must themselves be first introduced through language, as a purely
> psychological objection, clearly misses the core of this intuition. For that this
> happens is based solely in the wish to communicate, as a human, knowledge to
> other humans, which naturally can only happen in phonetic language
> [*Lautsprache*]. The language of symbols, however, is silence [*schweigen*]. Only
> the thinking subject itself would understand thoughts if the means of phonetic
> symbols [*Lautsymbolik*] were not used—which, in itself, is thinkable. Beings,
> whose language would be silence and whose communication would consist in
> the sign not of phonemes, but of things, could communicate logic without
> remainder in the manner of calculus. (S 2:110)

This passage develops the metaphorics of privative structure in mathematics;
mathematics restricts representation in language by excluding not only rhetorical symbols (like analogy) but also the semantically meaningful sounds
of human language. For Scholem, a pure, complete mathematical logic would

consist of a language of nonsemantic, nonphonetic, self-referential signs (*Zeichen*).[66] By emphasizing the cognitive potential of language, the *Referat* follows the epistemic transition charted out in Benjamin's "On Language as Such": Benjamin's analysis of Genesis displaces the epistemological reference point, moving from a visual-geometric frame of knowledge (as in Plato, Euclid, and, later, Kant) to language and the efficacy of the word, as in Adam's divine act of naming.[67] And yet this passage also takes Benjamin's thesis a step further, shifting the emphasis from a language dependent on its phonetic-semantic structure (*Lautsprache*) to symbolism (*Symbolik*) that speaks in a language of silence (*Schweigen*). The negativity of mathematical logic lies in this restriction of the rhetorical and phonetic features of language and its reliance, instead, on a language of "silence" composed of the syntactic grammar and logic of symbols. The *Referat* thus served as an index of and positioned itself in a deeper crisis of intuition in mathematics caused, as Volkert puts it, by the emergence of branches, such as mathematical logic, that function in algebraic and logical syntax, but evade visual-geometric intuition. Its proposed language of silent symbols also laid bare the central paradox that mathematical logic suggests: there may exist other and equally effective modes of language, even ones that do not function through the usual modes of representation available to language.

This paradox, however, is not a return to skepticism or nihilism, but rather the generative negativity that Scholem finds in the philosophy of mathematics. Expanding the idea of language by excluding "analogy" and "phonemes" but including "things" is the radical message the *Referat* delivers: it is "thinkable" that there may be "beings" who communicate through "silence," purely silent algebraic symbols. Scholem may have taken this idea from Paul Scheerbart's "asteroid novel" *Lesabéndio* (1913), which details the lives and aspirations of rubbery life forms who live on the asteroid Pallas. Scheerbart's novel presents an extraterrestrial cosmos filled with different forms of language—such as those of light and pressure.[68] Scholem's language of silence combines this multiplicity of languages in Scheerbart (and Benjamin) with the metaphorical horizon of the privative structure of mathematics and its constriction of language detailed in this chapter. Hence, if Benjamin's essay "On Language as Such" sought to overturn the bourgeois and mystical conception of language, then Scholem's *Referat* wanted to reveal the limits and alternatives to a "psychological" conception of language that takes intersubjective communication as its primary concern.[69] This alternative language set the stage for the duality of a "true" versus "decaying" tradition in Scholem's theorization of mysticism, as the mythical union of the mystic with God (*unio mystica*) remains as inaccessible to the human as the "great tautology" of "mathematics as such."[70] The paradox revealed by mathematical logic is thus a paradox that subtends not only mathematics but also traditions such as mysticism. We have a language to talk about the possibility of realizing the totality of mathematical logic and the divine realm sought and conveyed by mysticism. But this language is itself insufficient to complete these tasks in its form as a phonetic language and must undergo a radical transformation—such as in a mathematical logic that strips language of its rhetorical and phonetic-semantic register—to move past its inabilities.

The postulate in Scholem's *Referat* of a language of silence served as the point where the metaphorics of privative structure and the restriction of language in mathematics became operative as an aesthetic strategy in his work on lament. Indeed, the language of silence provided the leitmotif for his theorization of lament in the short text "On Lament and Lamentation" ("Über Klage und Klagelied," 1917) and his translations of the Book of Lamentations from the Hebrew Bible, both of which he composed directly following his intensive study of mathematical logic.[71] As a genre, lament (*qinah*) gives voice to and petitions God to account for situations and experiences of loss, deprivation, and pain, as captured in the Hebrew name for the Book of Lamentations in the incipit, *eikhah* ("how"; S 1:318). Although there are variants of lament in Jewish literature and thought from the Bible to medieval songs of lament, the translations of *eikhah* that Scholem completed in early 1918 are likely based on his version of the *Biblia Hebraica* (1913).[72] The *eikhah* lament the ineffable horrors of the destruction of Judah and the Temple, decry the enslavement of its people, their banishment and persecution in exile, as well as call on God for reconciliation and redemption. The salient feature of lament is that its content—the idea that God could let catastrophe befall the chosen people—exceeds the tools available to language to represent it in full: the extremity of these experience, like that of the Holocaust, lies beyond the limits of representation. These lamentations and lament in general offered Scholem not just a literary depiction of the Jewish historical experience of inexplicable privation, but also an opportunity to reconfigure language in order to instantiate this sense of lack—language's inability to represent such experiences—as a formal principle. Here lies the critical novelty of Scholem's theory of lament and his translations: they turn inexpressibility into an aesthetic strategy that, taking its cue from mathematical logic, mobilizes structural lack as a formal feature of poetic language to represent, in negative, the Jewish experience of privation recorded in the *eikhah*.

For Scholem, what allows lament to undertake the paradoxical task of representing experiences for which there is no language lies in the idea that lament occupies the liminal "border" between two regions of language: revelation and concealment ("des Verschwiegenen"; S 2:128).[73] As Scholem defines it, the fact that lament sits on this border region means that it neither reveals nor conceals its subject matter. Lament "reveals nothing, because the essence, which is reveled in it, has no content . . . and it conceals nothing, because its entire being [*Dasein*] is based on a revolution of silence" (128). Lament cannot fully reveal the loss and hardship of historical experience, because their extremity exceeds the limits of language. Lament mirrors mathematics for Scholem, in that it ventures to represents in language that which ultimately lies beyond language, instead of concealing it by not representing it at all. Indeed, as we saw in the absolute metaphors of mathematics, lament exhibits in Scholem's writing a negative if not paradoxical relationship to its subject matter, which exists beyond the comprehension of human mind and language, but which it, nonetheless, attempts to represent in poetic verse.

In "On Lament and Lamentation," this negative relationship between lament and its object becomes a matter of linguistic form. If the generative negativity in mathematics lies in a restriction of language, then lament takes this idea a step further: "Language in the configuration of lament annihilates itself [*vernichtet sich selbst*], and the language of lament itself is thus the language of annihilation" (S 2:129). What this passage means is that lament requires a special "configuration" of language, a "language of annihilation" that works against ("annihilates") language itself. Lament "annihilates" specifically "itself" because, as I will show shortly, it employs linguistic and literary strategies to oppose literary language. What exactly does lament annihilate? The answer not only recalls Scholem's mathematical Platonism but also distinguishes lament from mourning, in which images like the *memento mori*'s skull and bones intuitively and fully symbolize loss.[74] In the same way mathematics could not speak in "analogy" or "image," lament cannot be "symbolic" or "objective" (*gegenständlich*), because what it represents has "no content": the extremity of the experiences lamented makes them unavailable to the human mind to "symbolize" (128). Like mathematical logic, lament restricts representation in literary language by deliberately excluding symbolism; parallel to the privative structure of mathematics, lament constitutes "a fully autonomous order" cut off from the usual world of poetic symbolism. And yet this privation of language does not indicate lament's communicative impotence, but rather the creative potential of silence, which lies in grinding away at the means through which poetic language represents. Lament retains this positive ability to signify that there are experiences that cannot be represented in language, because "language has indeed sustained the fall of humankind, but silence," and with it lament, "has not" (133).

For Scholem, lament picks up where other forms of language fall short, because it redefines silence as more than just the absence of language. As Scholem calls mathematical logic a "revolution of logic," lamentation also draws on "a revolution of silence" (S 2:109, 128). The "revolution" lies in the rehabilitation of silence's epistemological and representative abilities:

> The teaching [*Die Lehre*] contains not only language, it contains in a particular way the language-less, the concealed, to which mourning belongs, as well. The teaching, which in lament is not expressed, not hinted at, but rather concealed, is silence itself. And, therefore as well, lament can take possession of any language: it is always the not-empty, but extinguished expression, in which its wanting-to-die [*Sterbenwollen*] and inability-to-die [*Nichtsterbenkönnen*] are connected. The expression of the innermost inexpressible [*Ausdruckslosen*], the language of silence is lament. (131)

Programmatically, the passage expands a concept of knowledge ("the teaching") beyond that which can be captured in language to include "the language-less" and the "concealed." But the passage here gets more specific: it delineates how lament produces an inverted and mute version of representation, not by trying to represent the "empty," but by presenting expression in the very process of being "extinguished." Lament, as Adorno later put it, serves as the "cipher" of the failed possibility of expression (A 7:178). To achieve this expressive annihilation of expression, Scholem's theorization of lament turns to his work on mathematical logic. Recall that in the *Referat's* discussion of mathematical logic, the completion of mathematical logic depended on a new language of "silence," the eschewal of human language, its "phonetic" structure and symbols. "On Lament and Lamentation" translates this restriction of language in mathematics into formal poetic strategies, such as meter: "the silent rhythm [*der schweigsame Rythmus*], the monotony is the only thing of lament that sticks: as the only thing, which is symbolic about lament—namely, a symbol of the state of being extinguished in the revolution of mourning" (S 2:132). Here Scholem plays on the meaning of the word *silence* (*Schweigen*) in German, which means both the state of being silent and the process of falling silent. Where mathematical logic rejects phonemes altogether (it is silent), lament repeats them until the meaning they impart to words begins to erode (it silences meaning). The "silent rhythm" produced in Scholem's translations thus not only enacts silence; by enacting, it also symbolizes on a poetic level the privation of language and, at the same time, the historical privation of the Jewish people that they lament.

Scholem's translations of Jewish lamentations employ a host of formal methods to wear away at the creation of meaning in poetic language. For instance, in "A Medieval Lamentation" ("Ein mittelalterliches Klagelied," 1919), Scholem's translation elongates single sentences over twenty lines.[75] Similarly, his translations of the *eikhah* from Hebrew into German emphasize such meaning-destroying monotony superficially by abandoning the traditional acrostic form as well as forgoing stanza breaks or verse numbers, as in the original *Biblia Hebraica*. These translations also accentuate the diminished stress of *eikhah's* 3:2 bicolon, characteristic of *qinah* meter, by splitting the original half-lines (three stressed words followed by two stressed words) into two or three new lines.[76] Take, for example, the sixth through ninth verses of the second lamentation:

> Er zerstörte wie den Garten seine Hütte,
> Verdarb seine Feste,
> Vergessen ließ Gott in Zion

Festfeier und Sabbat
Und verwarf in seiner Zorneswut
König und Priester.
Verschmäht hat Gott seinen Altar,
Verworfen sein Heiligtum,
Verschlossen in Feindeshand
Die Mauern ihrer Paläste.
Die Stimme erhoben sie im Hause Gottes
Wie am Tage der Festesfeier.
Gott dachte zu verderben
Die Mauer der Tochter Zion:
An legte er die Richtschnur,
Nicht wandte er seine Hand ab
Vom Verderben
Und gab Trauer über Mark und Mauer:
Sie sind verstört allzumal.

[Like a garden, he destroyed his huts,
Ruined his feasts,
God allowed Sabbaths and festivals to be
Forgotten in Zion
And dismissed in his indignant anger
Kings and priests.
God cast off his altar,
Discarded his temple,
Lost to the hands of the enemy,
The walls of her palaces.
They raised their voices in the house of God,
As on the day of a festival.
God thought to ruin the walls of
His daughter Zion:
He out stretched a line,
Did not restrain his hand,
From Ruination
And spread grief over rampart and wall:
They languished together.[77]]
(S 2:116)

Although such splitting preserves the meaning of sentences, it emphasizes the monotony of the unequal three ("*Vergessen* ließ *Gott* in *Zion*") followed by two stressed words ("*Festfeier* und *Sabbat*"). The breaking up of the half-lines defers the sentences' semantic impact, intensifying how the original forces readers to wait, in translation over the line break, to learn "what God let be forgotten" and whom he "dismissed in his anger" (lines 3 to 6). The cumulative effect over the five lamentations is the wearing away and deferral of meaning, which forces readers to read the poems aloud, not as semantic communication (as a *Lautsprache*), but rather as the enunciation of an unequal, symbol-less rhythm.[78] In their extinguishing of semantic meaning, their "silence," Scholem's lamentations signify on the level of form the inability to represent these events. The similarity and difference in mathematical logic and lament thus lies in that both restrict the symbolic function of language, but where mathematics strips language down to a syntax of arbitrary signs, lament wears away at language, leaving sounds that evince the erasure of meaning. Operative and creative in the philosophy of mathematics and lament for Scholem are these structures that abandon reference and semantics, indicating on the formal level the symbolization in negative of their own privation.

Scholem's theorization of lament turned the privative structure of mathematics into an aesthetics through privation. At the same time, mathematical logic's approach to negativity, its restriction of language, became in his translations of the lamentations from the Hebrew Bible a language that described historical privation, by enacting privation on the level of form. Lament, as Scholem writes, "only hints at the symbol" in its annihilation of symbolism and meaning (S 2:128). Benjamin, who would soon compose his own theory of translation, doubted the aesthetic success of Scholem's translations, as he wrote in response to reading Scholem's texts.[79] But Scholem's theory of lament and translations of the biblical lamentations nonetheless mark a significant point of conceptual transfer between mathematics and aesthetic theory. Both in theory and practice, Scholem's work on lament propose a set of critical techniques—a negative aesthetics—that represent the experience of diaspora, erasure, and loss through the removal of symbol, tireless monotony, the breaking of poetic verse, and an idea of silence not as inexpression, but rather as the erosion of expression and sense. In the context of Jewish exile and persecution, lament became for Scholem a means not only to test the limits of language but also to turn language into an expressive marker of its own limitations. It symbolized the fact that there are experiences whose extremity mean that they evade the usual representational strategies of human language, such as symbolization itself. Such a poetics, drawn from negativity in mathematics and applied in lament, gives voice to the experience of diaspora and erasure as such, if not also to the communities that experience such loss. The view of language presented by lament is thus critical, in that it recognizes experiences that language may more readily pass over because they exceed its limits. To the extent that lament expresses not the inexpressible, but rather inexpressibility itself, it intonates that silence and lack of meaning can be constitutive factors of poetic language.

Negative Aesthetics as History and Tradition

The privative structure that developed out of Scholem's work on the philosophy of mathematics yielded in his work on lament strategies for representing the experience of exile and loss—a negative aesthetics. In Scholem, negative mathematics reveals more about the nature of language and its potential uses in critical theory than Horkheimer and Adorno would suggest. Recall from chapter 1 that the same mathematical logic that Scholem studied in Jena threatened, in Horkheimer and Adorno's interpretation of logical positivism, to eliminate language and poetry as meaningless metaphysics, to render philosophy, as they said, "mute."[80] In contrast, negative mathematics offers new configurations of language, suggesting a poetic and even a critical dimension to such silence. By elucidating the seemingly paradoxical relationship between the absolute realm of mathematics and humans, negative mathematics provides critical theory with such a form of language, one that functions through restriction, taking the restriction of representation as a form of representation itself. Indeed, for Scholem, the deeper dimensions of representation revealed by negative mathematics were not a question of language alone but also, as shown in this chapter, a question of history and tradition.[81] The structure of lack that Scholem found in mathematics bears the possibility of transmitting histories and traditions that, like lament, function not despite, but because of privation. As in the Kabbalist, whose mystical tradition goes beyond the capacities of language as its medium of transmission, negative mathematics offers cultural criticism the idea that there may be histories and traditions that function through the very moments when historical representation and cultural transmission seem to break down.

Like Scholem's negative aesthetics, mathematics reveals a theory of history more open to the historical experiences of erasure and diaspora—experiences such as catastrophe, homelessness, and assimilation that challenge the limits of historical representation. Readers familiar with critical theory may recognize here a similarity to Benjamin's image of the "angel of history": while we attempt to represent history as a "series of events," the angel of history sees "a single catastrophe that relentlessly piles ruins upon ruins and hurls them before his feet."[82] Negative mathematics as a productive aesthetic theory in lament sheds new light on this theory of history, shifting the emphasis from the series of events that we call history and the singular catastrophe that Benjamin ironically calls the "progress of history" to the piles of ruins themselves. This theory of history not only challenges the notion of history as a narrative of progress but also tells history through these ruins, from the perspectives of exile and discontinuity, erasure and assimilation. In this regard, the writing of history would function along the lines of the privative structure of mathematics: it would dwell less on what remains of the historical record in language than it would attempt to construct history out of its silences, its lacks of meaning, and, as Michel Foucault puts it, "the irruption of events."[83] As is evident in Scholem's work on lament, negative mathematics offers critical perspectives on history and potential strategies for reconfiguring history to include—alongside narratives of what is representable and transmissible in language—indexes of events and experiences that the language of history and its narrative strategies cannot represent.

Consider briefly Scholem's own history of Jewish mysticism as an example of how such a theory may look in practice. Mathematics and mysticism both venture to represent phenomena that exceed the limits of the human mind and language, but mysticism, opposite of mathematics, depicts the mystic experience by employing—even, at points, to excess—the symbolic tools available to language. Texts such as *Major Trends in Jewish Mysticism* create a history out of mystical responses to Jewish persecution—persecution by the Church in the fourth century, the expulsion of Jews from Spain in 1492, if not also the "great cataclysm" of Scholem's own lifetime.[84] "The more sordid, pitiful, and cruel the fragment of historical reality allotted to the Jew amid the storms of exile," Scholem later wrote, "the deeper and more precise the symbolic meaning it assumed, and the more radiant became the Messianic hope which burst through it and transfigured it."[85] The privative structure of mathematics is at work here, only in inverse: where mathematics and lament fall silent, mysticism produces an excess of symbolic language as a marker of experiences and privations that lay beyond languages' limit. History configured around the negativity of mathematics would take into consideration the events and experiences, such as that of the mystic, otherwise not fully representable and transmissible in language. This theory of history could give voice to the voicelessness of diaspora and erasure by finding historical continuity in silence as well as the excess of symbols that covers up the silences of inexpressible experiences.

Furthermore, the privative structure of mathematics active in lament suggests a deeper dimension to the notion of historical continuity, bearing the possibility for traditions that continue despite historical rupture. As in the example of the Kabbalist that provided the starting point for this chapter, I refer here to tradition not only as the passing on of cultural practices and knowledge between generations but also the theoretical possibility of transmissibility as such. Take also the transmissibility of the Torah, the first five books of the Hebrew Bible and the rabbinic commentaries, which, in Scholem's words, starts to resemble the definitions and interpretations that attempt to express the negativity of mathematics:

> What is *Torah?* Under this term, I mean: (1) the principal, according to which the order of things is formed. Now, according to the perspective of Judaism, this principle is knowable too as the language of God [*die Sprache Gottes*] and, even in a specific manner, in the transmission of humans [*Überlieferung der Menschen*]. (It is here that the concept of tradition, as a corollary to that of the teaching, receives its unique meaning.) Within Judaism, to whom we owe the term, this implies (2) Torah as the integral, the epitome of religious transmission of Jewry, from the first days to the day of the Messiah.[86]

Although the divine laws given to Moses and recorded in the Torah are any-thing but arbitrary, this definition of tradition likewise separates the "lan-guage of God" and the "transmission of humans." Parallel to "mathematics as such," the first definition of Torah as "the language of God" exists inde-pendent of the accumulation (in Scholem's mathematical terms, "integral") of religious knowledge passed on across time. If this world of "mathematics as such" exists independent of our linguistic representations of it, then the divine definition of tradition would persist even when human interpretations were to come under threat from assimilation or catastrophe. As was the case with lament, the problematic transmission of the Torah would thus serve as the marker of the absolute division between knowledge passed on by humans and the divine word of God.

The negativity of mathematics affords a vocabulary to conceptualize such a theory of tradition that continues despite rupture. Indeed, for Scholem, the concept of continuity is deeply tied to mathematics; "is truth continu-ous," he writes in his so-called mathematical theory of truth, "is it always differentiable, that is, does everything have a concept and every truth an inner form?" (S 1:418). The passage invokes the mathematical ideas of con-tinuity, a property of a function that lacks gaps or breaks, and differentia-tion, which determines the direction and rate of change of a function for a specific value. A person's height, for example, is a continuous function of time, while the amount of money in my wallet is discontinuous, because it increases by a discrete amount when I am paid and decreases when I buy coffee. Traditionally, the continuity of a function for a certain value meant that one could also calculate the differential at that value. Yet developments in the nineteenth century on the syntactic-algebraic side of mathematics similar to mathematical logic challenged the intuitive relationship between continuity and differentiation.[87] For instance, Karl Weierstaß developed pathological, "monster" algebraic functions with a uniquely privative struc-ture: they are continuous everywhere but are differentiable nowhere.[88] In other words, this function would have no gaps, but we could not determine any second-order knowledge regarding its rate of change or direction—as Fenves explains regarding the "curve" of time: it "takes a sharp turn at *every* point."[89] This suggestion carries special significance for a theory of tradi-tion, in as much as it implies that contemporary crises in tradition do not entail a full break with tradition as such. As in lament, the privative structure of mathematics renders legible here how we can think of tradition as functioning not only in terms of the positive transmission of knowledge but also in the negative, as a symbol of tradition's independence from its own transmission. In this regard, tradition continues, even if contemporary observers may be unaware of, in mathematical terms, its direction or rate of change; "real" tradition persists, even as its "decay" seems to fade away. Such a theory of tradition would take such points of crisis, seeming erasure, and inexpressibility as not signs of discontinuity, but rather as the constitutive elements, like silence in lament, of tradition itself.

As with Scholem's negative poetics, these possibilities for theorizing history and tradition would take the privations of history and erasures of tradition as their generative spark—in writing a history of exile and a cultural tradition of lament. By turning history and tradition into an index of their own silences and erasures, they thus would encompass and afford theoretical room to historical experiences and cultural practices that rationalist discourse, majority cultures, and national, world-historical narratives may more readily marginalize or assimilate. Negative mathematics reveals these possibilities for aesthetic and cultural theory neither because it is somehow opposed to language, as Horkheimer and Adorno suggested, nor because it somehow calculates the trajectory of history or the limit of tradition. Instead, negative mathematics constitutes its own epistemological realm alongside history and mysticism, illuminating, based on its problematic relationship to language, the dark corners and hidden pathways of representation. But what if we allowed mathematics to speak with analogy and image—to work with the "integral" of tradition, the "continuity" and "derivative" of truth? What if we applied mathematics more directly to cultural criticism? What possibilities, if not also dangers, arise in using mathematics as an instrument of thought? It is to these questions that the next chapter turns.

THREE

Infinitesimal Calculus: Subjectivity, Motion, and Franz Rosenzweig's Messianism

Rational thought needs tools. From the collection of Aristotle's logic into an *organon*—meaning instrument or tool in Greek—to the contemporary academic's toolbox, philosophers employ tools such as the method of dialectics or the concept of subjectivity in order to analyze ideas, make arguments, and construct theories.[1] For Franz Rosenzweig, mathematics and, in particular, infinitesimal calculus provided thought with a new set of epistemic tools at the start of the twentieth century, when the instruments of reason inherited from the all-encompassing systems of German philosophy seemed no longer to apply to modern life. As discussed in chapter 1, however, Horkheimer and Adorno's equation linking mathematics, thought, and instrumentality drove the forces of control, oppression, and war, much like the "the printing press," "the cannon," and "the compass."[2] By the end of World War II, philosophy based on the symbols and operations of mathematics meant to them the instrumentalization of reason, which blindly confused the means of thinking with its ends and enabled enlightenment's return to myth and barbarism. When employed without reflection, mathematics turned nature into things to be calculated, exchanged, used, and, when necessary, destroyed. And yet, during World War I, Rosenzweig found tools in infinitesimal calculus that allowed him to grasp what other forms of knowledge and language could not express, and to refocus thought on the world of experience and the actions of the rational subject. Drawing on the synthesis of infinitude and finitude in infinitesimal calculus, this reorientation of thought set the terms for Rosenzweig's messianism and messianic epistemology, which placed the project of knowledge and the possibility of realizing emancipation in this world into the hands of the thinking individual.

We know Franz Rosenzweig today as a philosopher, educator, public intellectual, and cofounder of the Freies Jüdisches Lehrhaus in Frankfurt am Main. In 1921, Rosenzweig published *The Star of Redemption* (*Der Stern der Erlösung*), his main contribution to philosophy and theology. The primary achievement of the book, which Rosenzweig expanded in 1925 into a "messianic theory of knowledge," was twofold. First, it repositioned the living, thinking individual as the primary arbiter of philosophical and theological truth. Second, Rosenzweig's thinking showed how we can think of the truths established by the experience of the subject as equal to, if not more significant than those proved by mathematics. Drawing on the work of the Neo-Kantian Hermann Cohen, both of these claims in *The Star of Redemption* find a "guide" and "an *organon* of thought" in the mathematical determination of motion, namely infinitesimal calculus.[3] Infinitesimal calculus' approach to negativity, which hinges on the idea that infinitely small quantities mediate between nothingness and finite existence, provided Rosenzweig with a conceptual language to approach phenomenon (such as death) that remained incomprehensible to other systems of thought. Although Rosenzweig's interpreters have tended to pass over infinitesimal calculus as a mere "analogy" in his thought, this chapter reexamines the theoretical consequences of this generative negativity as it emerged in a metaphorics of subjectivity and motion in Rosenzweig's writings on mathematics.[4] Providing intellectual tools to the subject and redefining the concept of motion, infinitesimal calculus turned into a messianism in Rosenzweig's thought in which the work of the individual anticipates a messianic age that cannot be said to stand in a historical relationship to the present.

The ways of attending to negativity opened up by mathematics enabled Rosenzweig to find intellectual pathways through several philosophical and theological difficulties. The first of these was rendered legible for Rosenzweig not only because of World War I but also, and more immediately, by what he diagnosed as the disappearance of subjectivity in the philosophical tradition of German idealism after Kant and Hegel.[5] Although Rosenzweig's philosophy, as Benjamin Pollock shows, would not abandon idealism's systematic claim of "knowing the All," it maintained that this quest for the totality of knowledge had previously left the concerns of the subject, the individual philosopher untouched (R 2:21).[6] The second impasse addressed by negativity in mathematics was the claim—first presented to Rosenzweig by Eugen Rosenstock in the 1913 *Leipziger Nachtgespräch*—that Judaism no longer played a constitutive role in redeeming the world, in establishing the Kingdom of God on earth. The historical ascent of Christianity to political hegemony in Europe meant, Rosenstock argued, that it superseded Judaism as the path to redemption in the modern world, rendering the latter theologically obsolete.[7] For Rosenzweig, infinitesimal calculus lent a language to both issues that, through metaphorics of subjectivity and motion, made the world of experience intelligible to the individual and revealed the enduring messianic contributions of Judaism in the present. Combined, these two metaphorics drawn from mathematical thinking underpinned Rosenzweig's messianism, in which the infinite depth of Jewish religious experience and perpetual existence of the Jewish people anticipate and, thus, work toward the unreachable eternity of the Kingdom of God.

This emphasis on the redemptive force of individual action in Rosenzweig's thinking helps bring into focus the messianism still operative in cultural criticism today. As in the work of Benjamin and Adorno, such a messianism holds a view of history as potentially full, at every moment, of messianic interruption; "the elements of the ultimate state [*Endzustand*]" lie, as Walter Benjamin wrote in 1916, "deeply imbedded in every present [*in jeder Gegenwart*]."[8] This hope of the "emancipatory promise" of messianism remains active in cultural criticism, even as it becomes a messianism, in the words of Jacques Derrida, "without religion, a messianic without messianism."[9] Negative mathematics in Rosenzweig's thought helps us further pinpoint the influence of theology and, in particular, Jewish messianism on the critical project.

In traditional Jewish messianism, the Messiah remains absent in worldly life until the final redemption that marks the end of history; "the cataclysmic element remains otherworldly," Anson Rabinbach explains, "and consequentially makes redemption independent of either immanent historical 'forces' or personal experience of liberation."[10] In contrast, Rosenzweig's negative mathematics illuminates a messianism in which not only the individual moment but also the finite actions of individuals and groups in the here and now already contain a redemptive element, already work to reveal the world, in Adorno's words, "as it will appear one day in the messianic light."[11] Embedding the possibility of redemption in subjective action, I contend, makes up the critical contribution of negative mathematics in Rosenzweig's messianism and his messianic theory of knowledge. Both make emancipation, redemption, and the project of knowledge depend on the dynamic work of the thinking subject, even those who may belong to groups traditionally marginalized in philosophy, theology, and history. Both thus suggest that the individual, the critical theorist, is a primary worker in the creation of an emancipated society.

The generative negativity of mathematics, which helped Rosenzweig conceptually connect redemption with individual action, came in the form of the infinitely small quantity, also known as the differential. Given the technicalities of Rosenzweig's argument, let me first summarize what the differential is and how it functions in infinitesimal calculus. In the calculus developed by Leibniz and as the term is still used today, a *differential* refers to an infinitely small distance, often denoted as dx; Newton's calculus contained an analogous idea, namely the fluxion, which denotes an infinitely small instant of motion.[12] Differentials and fluxions enable the process of differentiation (fig. 3.1), which calculates the instantaneous rate of change of a curve at a specific point, and integration (fig. 3.2), which determines the area accumulated between a curve and the axis.[13] For differentiation, the ratio of the differentials PQ' and $P'Q'$ equates to the curve's rate of change at P as we make P' infinitely close to P; PP' is also called the tangent line, which touches the curve at only one point, P. For integration, the differential stands

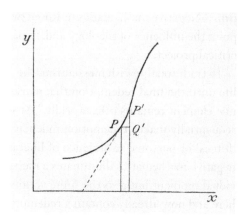

FIGURE 3.1. The ratio of differentials at point *P* is equivalent to the slope of the tangent line *PP′* that intersects the curve at point *P*, as depicted in Eduard Riecke's *Textbook for Experimental Physics* (1896).

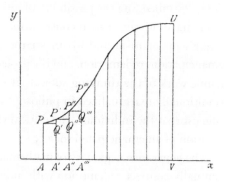

FIGURE 3.2. The integral equals the sum of an infinite number of infinitesimally narrow rectangles starting with *PAA′Q′*, as depicted in Riecke's *Textbook for Experimental Physics*.

in for the infinitely small widths of rectangles between *A* and *V*; as we make these widths infinitely small, the sum of an infinite number of rectangles approaches the area under the curve *U*.

In both examples, the differential serves as the conceptual bridge among nothingness, finitude, and infinity, generating the slope of the curve from infinitely small quantities and the area of the curve from the infinite sum of infinitely small widths. Through the metaphorics of subjectivity and motion, this mathematical approach to negativity translated into Rosenzweig's vision of redemption, which linked the eternity of redemption with the seeming infinitesimal actions of the individual. For critical theory today, Rosenzweig's linkage of human belief and action in the present with a messianic future suggests that the emancipatory potential of thought lies in the active engagement of critique in the here-and-now. As I intend to demonstrate, Rosenzweig's negative mathematics thus bears the possibility of a messianic form of epistemology, which accommodates the truths of mathematics as well as forms of knowledge that cannot be proved via mathematical and natural-scientific means—such as cultural criticism. Indeed, within this powerful yet circumscribed role, Rosenzweig's negative mathematics insinuates that critical theory should further reconsider the possibilities of instrumentality itself. By excluding negative mathematics, not to mention digital technologies based on mathematical processes, we otherwise risk forfeiting tools to explore and put into practice cultural concepts otherwise ungraspable in the critical project.

An Organ of Thought: Infinitesimal Calculus and the Metaphorics of Subjectivity

The initial conceptual steps toward this form of messianism and messianic theory of knowledge took place in the development of the metaphorics of subjectivity that Rosenzweig's early writings associate with infinitesimal calculus. For Rosenzweig, the individual was the philosophical element that the philosophy of idealism neglected. Idealism, associated with Kant, Hegel, and Schelling around 1800, also refers to the philosophical idea that reality and experience are, ultimately, products of the mind or spirit (e.g., *Geist* for Hegel).[14] If "philosophical reason" at the end of Hegel's philosophical system becomes "self-sufficient," if it "grasps all things and, in the end, grasps itself"—as Rosenzweig wrote in a letter from November 1916 that has become known as the "'*Urzelle*' [primordial cell] to the *Star of Redemption*"— then philosophy can grasp the human subject only in terms of its generality. What remains unexplored and unexplained is the unique individual, the philosopher: "I, a completely private subject, I forename and surname, I dust and ashes" (R 3:126). For Rosenzweig, infinitesimal calculus and contemporary pedagogical debates surrounding it took the "private subject" from the margins of philosophy and repositioned it at the center of thought. As another text from 1916, "*Volksschule* and *Reichsschule*," explores in the context of education (*Bildung*) reform, infinitesimal calculus provided the subject

with tools to know the protean foundations of the physical world and pedagogical methods to shape dynamic individuals. By grasping absence, infinitesimal calculus opened up the generative negativity of mathematics for
Rosenzweig, first coded in a metaphorics of subjectivity.

The metaphorical force of infinitesimal calculus in *"Volksschule* and
Reichsschule" hinges on the subject matter's place in contemporary debates
over mathematical education in humanistic education and in contrast to
other methods of mathematical pedagogy, such as those based on geometry.[15] Indeed, geometry had long served as a methodological paradigm (the
more geometrico, "in the manner of geometry") in philosophy as a system of
fully ordered reasoning; Euclid's geometry begins with well-defined axioms
and constructs its theorems based on these axioms via the rules of logical
inference.[16] The exemplarity of geometry informed neo-humanist, mathematical pedagogy in the nineteenth and early twentieth century in Germany
as well, in as much as mathematics taught through geometry constituted "a
self-contained set of propositions whose ideal harmony reflected the can-
ons of neohumanist aesthetics."[17] During Rosenzweig's lifetime and amidst
the mathematical developments detailed in chapter 2, the pedagogical focus
on the geometric ideality of mathematics increasingly came under scrutiny, as reformers such as Felix Klein argued that teaching such logical
precision was unnecessary for a nonspecialist mathematics education. In
contrast, Klein advocated "the strengthening of intuition [*Anschauungsver-
mögen*] and the education toward a propensity for functional thinking."[18] The
emphasis on mathematics as intuitive and goal-driven thought underpinned
the official reform recommendations for mathematical education at German
institutes of higher learning, known as the Meraner Reforms, announced
in 1905—the year Rosenzweig began his medical studies in Göttingen,
where, coincidentally, Klein was a professor of mathematics.

Not only for mathematical reformers but also for Rosenzweig's essay
"Volksschule and *Reichsschule*,"* infinitesimal calculus embodied this intuitive
and practical dimension of mathematics. According to Klein, infinitesimal
calculus should thus be a mandatory element of mathematical education and
should be taught, especially for modern professions requiring technical ex-
pertise such as engineering and medicine, as a "mathematics of approxima-
tion" (*Approximationsmathematik*). With a focus on concepts and practical

applications, this method of teaching infinitesimal calculus complemented an emphasis on algorithmic structure and precision, "mathematics of precision" (*Präzisionsmathematik*), designed for students of mathematics.[19] Although Rosenzweig was most likely unaware of these debates, the same intellectual dynamic motivates "*Volksschule* and *Reichsschule*," where the utility of infinitesimal calculus is indispensable for modern education and the subjects it produces. Indeed, the essay argues that mathematical education should be rearranged "so that differential and integral calculus can enter into instruction as early as possible" (R 3:392). At the "Reich's School" (*Reichsschule*), designed for university-bound pupils and equivalent to today's *Gymnasium*, infinitesimal calculus is thus the key, according to Rosenzweig, to reforming the natural sciences, which serve as one of three educational "organs" alongside the study of history and language. While this tripartite structure of "organs" resurfaces in *The Star of Redemption*, it guides the essay's emphasis on the modernization of education rendered necessary by World War I—not as "ruthless utilitarian education" (*Nützlichkeitserziehung*), but rather a recommitment to humanistic *Bildung*, the cultivation of such "organs" as they apply for "today's human" (R 3:383 and 374).[20] In other words, reorienting mathematical education around infinitesimal calculus in "*Volksschule* and *Reichsschule*" meant repositioning thought around the needs of the subject in a rapidly modernizing world.

For Rosenzweig, infinitesimal calculus accomplished this reorientation by offering a mathematical alternative to what he saw as the rigidity and idealism of geometry and geometry-based mathematics education. Beyond drawing on introductory textbooks in infinitesimal calculus that responded to Klein's calls for reforms, "*Volksschule* and *Reichsschule*" channels cultural perceptions of geometry as a paradigm not of proper reasoning, but rather of rigidity cut off from the world of experience.[21] In the words of Arthur Schopenhauer, geometry is a static form of thought, devoted to casting "easily accessible intuitive evidentness willfully aside and replacing it with logical evidentness," akin "to someone cutting off his legs so that he can go on crutches."[22] The apparent separation of geometry and the world of experience constitutes the major problem with contemporary mathematical education in "*Volksschule* and *Reichsschule*," which functions along the lines of Plato's dialog *Meno*. In *Meno*, Rosenzweig claims:

a theorem is drawn out of a slave, a theorem previously totally unknown to him; he is a mathematician "sans le savoir"; this is supposed to show how mathematical knowledge lies beyond experience [*das Übererfahrungsmäßige*] and how the Platonic Idea relates to experience; geometrical truth was in a certain sense simultaneously a subcase and a complement to the truth of the idea. Plato was justified to set above the door to his academy a motto prohibiting the entrance of "those without geometry" [*Geometrielose*], for just as geometry is the teaching [*Lehre*] of fixed forms, mathematics is the teaching of "things," so too Plato's idea stands behind the confusion of relations and the flow of appearances as a resting [*ruhenden*] relation-less form.[23] (R 3:389)

This passage sets up the counterexample against which the metaphorics of subjectivity and motion that Rosenzweig associates with infinitesimal calculus intervene. On the one hand, the passage codes the subject educated in geometry, according to the idea that "mathematical knowledge lies beyond experience," as the opposite of an active, engaged, and enlightened subjectivity: "a slave," who knows passively, "sans le savoir." On the other hand, the passage associates geometry with rigidity and detachment from the world of experience: "the geometry of Euclid" studies "fixed forms," "stands behind the confusion of relations," and constitutes "a resting, relation-less form" (389). Euclid's geometry exemplified for Rosenzweig a deeply limited vision of mathematical education and thought, in as much as its definitions and theorems make claims only about stationary, motionless shapes (such as lines, circles, and triangles) and adhere to—indeed, set the standard of—a strict logical construction.

The stagnation embodied by the prominence of geometry in contemporary mathematical education ran parallel to what Rosenzweig saw as the moribund state of philosophy around 1900, against which he purposefully aligned his reorientation of thought. Discontent regarding the institution of philosophy was widespread among intellectuals in the first few decades of the twentieth century, an "exodus" Margarete Susman characterized as a rebellion against "every kind of philosophy in terms of pure thought."[24] As formulated in the "*Urzelle*" letter, this rebellion crystalized most prominently in Rosenzweig's dissatisfaction with and rejection of German idealism, which ultimately, he writes, takes "the form of logical knowing A = A" (in symbols borrowed from Fichte).[25] In "*Volksschule* and

Reichsschule," however, it is Euclid's geometry—and not the German philosophers of the previous centuries—that exemplifies this type of totalizing, abstract reason as "the first and, if you will, purest classical, unsoiled form of European idealism" (R 3:389). Euclidean geometry constructs its objects solely in the mind of the thinker via definitions and logical deduction; it is thus "pure" and "unsoiled," because it does not, in contrast, for instance, to Kant's transcendental idealism, derive knowledge from empirical experience. In Euclid's system of knowledge, ideal rules determine what we call experience, which conforms to and, thus, serve as a confirmation of these rules. This point is important, because it distinguishes Rosenzweig's negative mathematics from the Platonic independence of mathematics that led Scholem to his version of negative aesthetics (see chapter 2). For Rosenzweig, the systematic exclusion of experience from both geometry and knowledge rendered legible the need for a philosophy that took the world accessible to the individual into account.

In "*Volksschule* and *Reichsschule*," infinitesimal calculus not only appears as a solution to reforming mathematical education but also provides the conceptual tools to accomplish this reorientation of thought. The essay juxtaposes the "Euclidean petrification [*Erstarrung*] of mathematical thinking" with the potential contribution of mathematics to *Bildung* as a whole: the ability to determine motion mathematically. Rosenzweig explains:

> Hence, at the end of an epoch, the greatest philosopher and greatest researcher of nature found the process, which once and for all redeemed [*erlöste*] nature for the mathematician from its rigid Euclidean sleep [*Starrschlaf*] in space and made the modern concept of nature accessible for mathematical thinking. In that we learned to think of the curve as "arising" [*entstehen*] from the line, the quantity "arising" out that which has no quantity (the "infinitely small"), we had discovered the method by which we could mathematically grasp motion as the primordial phenomenon [*Urphänomen*] and rest only as a limit-post [*Grenzpfahl*]. Namely, since the limit-post here was construed as the origin of real appearances, the "primordial phenomenon," mathematics found its accustomed point of departure secured in pure space, but only as a point of departure, in order to catch a glimpse, with the momentum of the new method, of the ungraspable and fugitive primordial phenomenon of the spatial-temporal world. (R 3:390–391)

Essentially, this passage describes the operation of differentiation; in the terms laid out above, the differential ("the infinitely small") determines tangent "lines," which we can think of as constituting a "curve" at every single point. In Rosenzweig's language, we notice a distinct association of infinitesimal calculus with a metaphorics of "motion" (*Bewegung*) in terms such as "arising," "momentum," and "fugitive" that contrast the "rigid Euclidean sleep" of geometry—we will return to these ideas in the next section. For now, I would like to focus on the metaphorics of subjectivity, which here is linked to infinitesimal calculus. Where "*Volksschule* and *Reichsschule*" associates geometry with Plato's slave, it speaks of differentiation in terms of the early-modern period's "greatest philosopher" (Leibniz) and "greatest researcher of nature" (Newton). Indeed, the historical narrative of intellectual progress told by the essay hinges on the "great" subjects of history, such as Leibniz, Newton, and Goethe, and their ideas, such as Goethe's "primordial phenomenon" (*Urphänomen*).[26]

What is significant about this metaphorics of subjectivity associated with infinitesimal calculus is how it helps mediate negativity on two levels in this passage and in "*Volksschule* and *Reichsschule*" as a whole. Mathematics offers the subject an "organ" (related to the Greek term *organon* meaning "tool" and "instrument") to know that which cannot be known otherwise, in Rosenzweig's terms, the "disassociate mass of the world." According to the previous excerpt, the discovery of infinitesimal calculus made "the modern concept of nature accessible for mathematical thinking," infinitesimal calculus grasps "motion as the primordial phenomenon," and provides a method "to catch a glimpse" of this otherwise "ungraspable and fugitive" motion (R 3:383). With differentiation, to follow Rosenzweig's example, mathematics generates lines and curves out of absence and negativity, "that which has no quantity." In this regard, infinitesimal calculus affords the subject powerful tools for dealing with negativity: the individual—the "mathematician," Leibniz and Newton, and the student in the *Reichsschule*—generates knowledge about the "primordial phenomenon," "the origin of real appearances," which, the excerpt suggests, consists of amorphous matter in constant motion. This is the generative negativity that mathematics provides for Rosenzweig, the ability to grasp what lay beyond the limits of other forms of knowledge; "mathematics," Rosenzweig later wrote, "is after all the language before revelation" that renders legible to the individual the created but not yet revealed and ordered world.[27] For Rosenzweig, mathematics provided tools to the thinking subject to generate knowledge from negativity, from the primordial, disassociate mass of motion that constitutes the world and remains concealed to other forms of knowledge, such as history and language.

Here we start to see the contours of the relationship among mathematics, negativity, and metaphor that governed Rosenzweig's contribution to the project of negative mathematics. For Rosenzweig, infinitesimal calculus served as an "organ," it offered a "language" to decipher the elements that made up the world of experience that other forms of knowledge and other languages could not grasp. This created but yet-to-be-revealed world was not mathematical per se. Instead, mathematics provided Rosenzweig with a means to work productively with this negativity—namely, a transition point between "quantity" and "that which has no quantity." As I will argue in the following section, the generative tool of mathematics was differentiation, which generated the knowledge of motion, as the real state of the world. For Rosenzweig, mathematics did not consist of absolute metaphors (see chapter 2), because we intuit—indeed, we experience—the physical world it allows us to describe. Instead, as John H. Smith suggests, we can think of metaphor in Rosenzweig's negative mathematics in the "strong, Aristotelian sense," used, in Aristotle's words, "in naming something that does not have a proper name of its own."[28] Here, mathematics provides a language for the subject to understand, at least in the physical world, what otherwise lays beyond language. In Rosenzweig's later work, infinitesimal calculus functions analogously providing names to the complexities of time and history as to include the theological contribution of Judaism.

One final point about mathematics and subjectivity: the repositioning of thinking around the individual achieved by infinitesimal calculus made subjects for Rosenzweig active creators of knowledge instead of passive agents erased by the knowledge they produce. In *"Volksschule* and *Reichsschule,"* the goal of *Bildung* lies not in the simple once-and-for-all possession of the "organs" of mathematics, language, and history, but rather in the ability to learn from and cultivate further these organs throughout the pupil's life. Again, geometry provides the counterexample: "I don't remember the details well enough anymore," Rosenzweig wrote to his mother, describing his own experience in geometry courses at the *Gymnasium*, "but I know how much value was placed on making things we had already learned routine." Such routine was, Rosenzweig continues, an incorrect pedagogical principle—"to make things routine by repeating them; things become routine by moving forward. When I'm required to employ things today that I learned yesterday as a way to new discoveries, that's when things become routine, not by repeating them countless times" (R 1.1:268).[29] This passage argues against a pedagogy, if not an epistemology, based on the slavish "repetition" of that which students already know, signified by geometry and reflected symbolically in the statement "A=A." Such instruction neglected the individual student in favor of the subject matter as a model of infallible reasoning. Similar to the reorientation advocated by reformers of mathematical education such as Klein, Rosenzweig's proposed pedagogy sought to reverse this stagnation surrounding the student by offering a dynamic, "process"; learning was of value, for Rosenzweig, only when it involves the action of the student in real time, the employment "today" of that which was acquired "yesterday." One notices here how mathematics suggests another way to deal with negativity, privileging motion over stasis (the absence of motion) in education and, ultimately, epistemology.

Reorienting knowledge around the subject was a key aspect of Rosenzweig's messianism and messianic epistemology. In contrast to a theory of knowledge that upheld the eternal validity of logical constructions, mathematics in the form of infinitesimal calculus provided tools to the subject, emphasizing, for Rosenzweig, the individual's contribution to the creation of knowledge. To be sure, Rosenzweig's dynamic, knowing subject armed with the tools of mathematics appears deeply troublesome from the perspective of Horkheimer and Adorno's critical theory. For Horkheimer and Adorno, subjectivity and instrumentalization were essential elements of domination and control; Odysseus—tied to the mast, yearning for the sirens while his men slavishly work to propel his boat forward—embodies the oppressive self-restriction of this form of subjectivity.[30] But what the tools of infinitesimal calculus and the metaphorics of subjectivity nonetheless rendered legible for Rosenzweig was the irreducible role of the individual thinker—the role not of Odysseus but rather of the critical theorist—in the ongoing project of knowledge. Reorienting philosophy around the diverse epistemological contributions made by a plurality of subjects would mean a more capacious and, indeed, more comprehensive vision of knowledge, which was precisely what the concept of motion in Rosenzweig's messianism achieves.

Quanta Continua: *Time, the Metaphorics of Motion, and the Necessity of Judaism*

By enabling the individual to capture mathematically the "primordial phenomena" of motion, the metaphorics of subjectivity already implied a metaphorics of motion in Rosenzweig's writing. Infinitesimal calculus began to reposition the individual, living philosopher at the center of the project of knowledge, but it also responded for Rosenzweig to the idea that Judaism no longer seemed to play a historical role in moving the world closer to redemption, the realization of the Kingdom of God on earth. This viewpoint and Rosenzweig's conviction of its falseness emerged out of the 1913 *Leipziger Nachtgespräch* among Rosenzweig, Rosenstock, and Rudolf Ehrenberg. Creating a redeemed world, as Rosenstock demonstrated to Rosenzweig, remained a historical option open only to Christianity, as Judaism lost worldly power with the destruction of the Second Temple and Christianity gained it with the Christianization of the Rome; this perspective left "no room for Judaism," as Rosenzweig later wrote, "in this world" (R 1.1:134).[31] Infinitesimal calculus, and what it revealed about motion, rest, and time, suggested otherwise. By revealing the primacy of motion over rest in the physical world, I argue, the implications of the discovery of infinitesimal calculus allowed Rosenzweig to reorient knowledge around the world of experience instead of the ideality of rest. At the same time, infinitesimal calculus's approach to negativity and absence helped rearticulate an otherwise inexpressible relationship between the here and now and redemption, in which Jews, too, could take part. As a metaphorics of motion, infinitesimal calculus showed Rosenzweig the deep structure of temporality, providing the rhetorical tools to render legible the necessity of Judaism as an eternal indication to Christianity of the incomplete and, hence, infinite path leading to redemption.

In my analysis of mathematics and the metaphorics of motion, I do not wish to suggest that infinitesimal calculus provided Rosenzweig's only conceptual tool in reinstating Judaism into the framework of redemption. His negotiation of this topic and his ongoing disputation with Rosenstock drew on myriad concepts from theology, philosophy, history, and literature.[32] However, Rosenzweig availed himself of infinitesimal calculus in particular, as in one revealing moment from his correspondence with Rosenstock in 1916, when theological language otherwise failed:

> Beginning and end, I could put it so, are the same for you [Christians] as they are for us [Jews]; with Newton to help with the analogy [*als Gleichnishelfer*]: the curve as "just arising" and as "just disappearing" has the same equation for you and for us. And, you know, that the entire curve can be determined by such differential quotients, but, in the course of the curve, you and we choose different points to describe it—and that is our difference. With Moriah and Golgotha, you correctly grasp this difference. (R 1.1:283–284)

Rosenzweig's goal here is to counter Rosenstock's claim that, with the fall of the Second Temple, Judaism had become detached from the world and, hence, from a trajectory of world history that ends in redemption. Similar to the dynamism ascribed to infinitesimal calculus in "*Volksschule* and *Reichsschule*," the analogy works by recoding negativity—the unknowable and, in Rosenzweig's words, "inexpressible" trajectory of time—in terms of motion, an "arising," and "disappearing" curve.[33] The rhetorical link between temporality and infinitesimal calculus makes intuitive sense, because infinitesimal calculus is the branch of mathematics used to calculate motion (change of place in time). In Rosenzweig's analogy, if we think of the totality of time as the "course of the curve," whose "beginning" is creation and "end" is redemption, then infinitesimal calculus provides tools with which we are able to describe the path of this curve at different points—namely, the "points" of Judaism and Christianity. In terms of Rosenzweig and Rosenstock's conversation, these points corresponded to central moments of sacrifice from the Jewish and Christian traditions. Accordingly, both represent the religions' equal but divergent demonstrations of faith: in Judaism, as Abraham proved his willingness to sacrifice his only son Isaac and the future of his people on Mount Moriah, while Jesus sacrificed himself for the future of his people with his crucifixion on Mount Golgotha.

Associating temporality, motion, and infinitesimal calculus, this analogy provides Rosenzweig with a vocabulary to articulate the idea that there could be more than one way to relate to and work toward the redemption of the world. As the term "differential quotients" suggests, the key to the analogy lies in the process of differentiation, which in the calculi of Newton and Leibniz determines the instantaneous motion of a curve at a particular point. Take, for example, the explanation of differentiation offered in Newton's *On the Quadrature of Curves* (*De Quadratura Curvarum*, 1704), a translation of which was in Rosenzweig's personal library.[34] Although the passage in Rosenzweig's letter references Leibniz's calculus with the term "differential quotient," the analogy also builds on Newton's redefinition of the idea of motion. For Newton, "curves" did not consist of "extremely small parts," to which critics of Leibniz's calculus objected, but instead were "described and, in the act of describing, generated [*erzeugt*], . . . through the continuous motion of points."[35] As with his terminology, Rosenzweig uses imagery that is more intuitive than historically or mathematically precise. His letter to Rosenstock describes the curve as seen in Newton's illustration, as something that "arises" and "disappears" as the point C "generates" the curve in the path of C's continuous motion as a point (fig. 3.3).[36] Multiple points along the curve of time—both Judaism's or Christianity's moment of sacrifice—tell us the direction of the curve, in as much as differentiation allows us to calculate its instantaneous rate of change as the ratio (in Leibniz and Rosenzweig's terms, the "differential quotient") of the curve's vertical displacement (cE) to its horizontal displacement (CE) as c moves infinitely close to C. Conversely, the ratios of cE to CE enable us to reconstruct a formula for "the entire curve" in terms of the "differential quotients" of different points on the curve.[37] Translated back into Rosenzweig's analogy: despite their different ways of engaging in the world, both Christianity and Judaism provide insight into the nature of their shared temporal trajectory. Infinitesimal calculus put into words the idea that, even as Christians work toward redemption in the world, Jews also help determine the "entire curve" leading out of creation and into redemption.

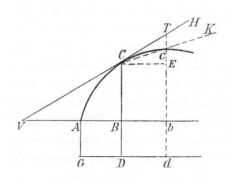

FIGURE 3.3. The differential quotient is the ratio of *cE* to *CE* as *c* moves infinitely close to *C*, as depicted in Gerhard Kowalewski's 1908 translation of Isaac Newton's *De Quadratura Curvarum* (1704).

And yet the cultural resonance of motion in the early twentieth century suggests a deeper interpretative layer to Rosenzweig's choice of Newton "to help with the analogy" and his depiction of the metaphysics of history as the motion of a curve. During Rosenzweig's lifetime, the idea of motion, speed, and dynamism set the terms for modernist artists, thinkers, and movements across Europe.[38] For example, Henri Bergson's concept of the *élan vital* expressed the dynamic principle active in all forms of biological life.[39] Recall, too, from the previous section that, for Rosenzweig, "motion" (*Bewegung*) constituted the "primordial phenomenon" that infinitesimal calculus rendered legible to the subject. Within Rosenzweig's intellectual universe, the concepts of motion, subjectivity, and mathematics intersected in the work of Houston Stewart Chamberlain, whose *Immanuel Kant* (1905) Rosenzweig cited as his source on the mathematics presented in *"Volksschule* and *Reichsschule."*[40] At first, this line of intellectual influence may seem dubious, given the *völkisch* and deeply anti-Semitic objectives of Chamberlain's thought.[41] But Chamberlain offered a worldview that was animated by terms such as "dynamism" (*Dynamik*), "motion" (*Bewegung*), and "force" (*Kraft*). For Chamberlain, these ideas underpinned the Western European ideal of educated, bourgeois *Kultur,* which Kant epitomized as the great "personality"—as (in terms that reappear in Rosenzweig's depiction of infinitesimal calculus) "a *punctum evanescens*, a continual arising and disappearing."[42] Similar to the description of the natural world in *"Volksschule* and *Reichsschule,"* this worldview took motion as its point of departure: "What is singularly decisive since Galileo and Descartes is that the symbolism of *motion* provides the foundation for every science of nature [*Wissenschaft der Natur*], that is dynamism, the presentation of force," Chamberlain writes; "rest is only a phase of motion and can only improperly be assumed with reference to the relationship between certain equally moving bodies. What is given is motion, absolute rest would be the nothing."[43] This worldview added to the metaphorics of motion a fundamental shift in how to approach the physical world, not from the starting point of "absolute rest," but instead from the foundation of motion, the dynamic world in which one lived.

At this point, we notice how Rosenzweig's metaphorics of motion pick up and expand the metaphorics of subjectivity: the tools that reorient thought around the individual offer a dynamic worldview that accommodates the distinct Jewish contribution to the project of redemption. Indeed, Rosenzweig's return to infinitesimal calculus in the middle of World War I and the resulting worldview based on motion stemmed, as he relates in a letter, from his work in theology; it was the "Thomasian proof of God out of the concept of motion," in reference to Thomas Aquinas, which rekindled his interest in mathematics leading up to "*Volksschule* and *Reichsschule*."[44] Here, infinitesimal calculus provided a key transition away not only from the idealism of Euclid's geometry but also what Rosenzweig saw as the ideal world of Aristotelian physics. He explains:

> Infinitesimal mechanics is the first solution to Zeno's paradoxes, which, for its part, was based on antique mechanics that starts with rest [*Ruhe*] and attempts to explain motion [*Bewegung*]. Newton's "first law," according to which rest is a limit-case [*Grenzfall*] of motion, helps appearance [*Augenschein*], which knows only motion and no rest, claim its due authority in *nature* (different than in logic). Newton's "first law" is the dethroning of Aristotelian mechanics (and, at the same time, of the doctrine of creation as a *theory of nature*). . . . In contrast, modern infinitesimal mechanics presupposes motion (that is, space and time) and thus can obtain rest as a limit-post [*Grenzpfahl*] of motion through differentiation.[45] (R 3:84)

This passage is dense with references and shorthand that beg to be broken down: In "Aristotelian mechanics," rest served as the ideal physical state of the earth knowable through the static figures of Euclid's geometry; things moved only when acted upon by other things. According to Rosenzweig,

Newton's new physics turned this theory of the physical world on its head. After Newton and his "infinitesimal mechanics," motion became the starting point of analysis, subject to measurement; rest became a special case [*Grenzfall*] of motion with a velocity of zero (its, to use Rosenzweig's pun, limit-post [*Grenzpfahl*]). The process of differentiation allows us to calculate the direction and rate of change of motion—zero, in the special case of rest. What is significant about this passage is how it announces two epistemological shifts associated with the new concept of motion inaugurated by infinitesimal calculus—both crucial for Rosenzweig's messianism. The first makes the world of experience the site of epistemology, whereas the second upholds that, even if Judaism may no longer play a major historical role in the world, it still occupies a constituent place on the "curve" of time, stretching from creation to redemption.

The first and explicit shift associated with the metaphorics of motion refocused knowledge away from the ideality of "absolute rest" in Aristotle's physics to the "appearance" of motion (*Augenschein*) in Newton's. This shift hinged on developing concepts of motion in the physical world. For Aristotle, motion constituted a form of change: "the actuality of that which exists potentially, in so far as it is potentially this actuality."[46] Motion is, in other words, the transition of a potentiality (I have the potential to walk to the door) to an actuality (I walk to the door). Accordingly, the motion of an object was determined by an object's essence: the heavenly bodies move in perfect circular fashion, in as much as a prime mover acts upon them to move as such. The essence of an object like a stone, however, is not to move unless a foreign force is added to it; without the continued application of force, the stone will reassume its natural state of rest.[47] This view is a "doctrine of creation," in as much as it tells a story of "nature" as emanating from rest, out of absolute nothingness. For Rosenzweig, Newton's physics "dethrones" Aristotelean mechanics because it explains the physical world by starting with "appearance," the world as it appears to the observer—in the German, literally "what appears to the eye" (*Augenschein*). Instead of starting with rest, the first of the three "Axioms, or Laws of Motion" proposed in *Philosophiæ Naturalis Principia Mathematica* (1687) upheld that a moving rock stays in motion unless another force (friction, for instance, caused by gravity) brings it to rest; the rock "perseveres in its state of being at rest or of moving" unless "it is compelled to change its state by forces impressed."[48] For Rosenzweig, this change was the epochal shift initiated by infinitesimal calculus and signified by the metaphorics of motion, in which the hitherto "fugitive" world of "appearances" became knowable through mathematics. This shift was significant, because it oriented knowledge around not an ideal world of static figures at absolute rest, but rather the dynamic, physical world as experienced by the individual knower. Emphasizing motion over rest meant, at least for Rosenzweig, a theory of knowledge that captured the messy, complicated, and constantly changing world in which we live and in which we have the potential to act and engage.

The second epistemological shift brought about by Newton corresponded to a transformation in how one thinks about the mathematical continuum. The previously quoted passage hints at this transition in notions about the continuum by calling infinitesimal mechanics the "first solution to Zeno's paradoxes." The mathematical continuum refers to the set of real numbers, which likewise constitute the real number line, and their properties, such as their arrangement and density. For Aristotle and, later, Thomas Aquinas, the continuum was a continuous magnitude—such as space, time, or a line—and started and ended in an "atom," a "now," or a "point," but was not made up of individual atoms, nows, or points itself.[49] As Euclid writes in *Elements*, points can only be "the ends of lines" and not their constitutive parts, because points (and, analogously, atoms and nows) are noncontinuous in that they lack extension, whereas a line (as well as space and time) is continuous.[50] This concept of motion led to one of Zeno's well-known paradoxes: an arrow in flight is in continuous motion, but, taken at any one instance, appears to be at rest, which contradicts the idea that motion does not contain moments of rest.[51] The Newtonian worldview eliminated these conceptual tension between motion and rest. As we saw with Newton's *On the Quadrature of Curves*, a continuous magnitude like a line is not bookended by points, but rather "generated" by a point as it moves continuously through space. This concept of motion "solves" Zeno's paradox by reframing the arrow's "rest" as only its instantaneous "appearance" at any one point along the arrow's path. For Rosenzweig, mathematics showed that a continuous magnitude, such as the "curve" of time, was not an impenetrable concept bookended by two points, but rather the product of a continuous "now" as it moves from a "beginning" in creation to an "end" in redemption. As intuitive as this concept of the continuum may seem today, this change in perspective was significant for Rosenzweig, because it meant that the continuum included all points drawn out as the present moves from the past into the future.

In particular, the idea that the continuum contains all its constitutive points revealed for Rosenzweig the possibilities not only that we can generate knowledge about a continuum like time, but also that all such points are indeed necessary to constitute the continuum as a whole. The new Newtonian continuum did not consist of an aggregate of indivisible, infinitesimal points or "infinitely small" moments; rather, it entailed two dimensions of infinitude as continuous motion: continuous without end, but also infinitely continuous, without gaps, at any one moment—an infinite expanse, so to say, of infinite density. While this idea was the focus of mathematical research in nineteenth century, it is already present in Kant, who writes, in reference to Newton that "space and time are *quanta continua . . .* [o]ne refers to quantities of this kind as flowing." As continuous magnitudes for Kant, "space consists exclusively of spaces, time of times. Points and instances [*Augenblicke*] are only limits [*Grenzen*], i.e., mere places of their limitation [*Einschränkung*]."[52] Here, as in Rosenzweig's use of the term, the idea of limit is more heuristic than algebraically precise, but it means that "time" is infinitely dense with "times" of which an "instance" is the smallest perceptible unit. Motion with space and time as "flowing quantities" implied that, for Rosenzweig through the Newton analogy, we can generate knowledge about time like we can with motion, namely through differentiation: on the one hand, differentiation calculates the direction and rate of change of time at any particular time and, on the other hand, we can describe the curve of time from any particular time via its differentials. As in Kant's example, the Newtonian concept of the continuum also implied that the continuum of time necessitates all of its constitutive "times," not just specifically significant or idealized ones, such as the endpoints. In other words, mathematics suggested that time, as a continuum leading from creation and ending in redemption, consisted of not only world-historical events but also of the infinity of other points that lie in between these moments and fill out and make up the dually infinite continuum.

For Rosenzweig, this more spacious concept of the continuum opened up new ways to think about the place of Judaism in the schema of redemption. Take another example in which Rosenzweig compares Judaism and Christianity from 1918, which invokes the Newtonian conception of the continuum even though it lacks a rhetorical emphasis on motion:

> You also see in that mathematical analogy how Judaism appears from the perspective of Christianity. The irrational number—what does it mean then for the rational number? For rational numbers, infinity would be the eternally unreachable limit from above or below, something eternally *improbable*, although surely and forever true. Precisely at the irrational number this limit of the world of rational numbers strikes every single one of its points, physically, numerically, presently, and redeems [*erlöst*] the world of rational numbers from the "linear" abstraction and uncertainty of its one-dimensionality into a "spatial" entirety and, hence, to a self-assured reality. (R 1.1:561–562)

Hidden behind the equation of Judaism with irrational numbers and Christianity with rational numbers is a reference to the makeup of the continuum; real numbers are either "rational," in that they can be expressed by a "ratio" of integers, or "irrational," such as π, in that they cannot. As Rosenzweig explains in full in the letter, the passage associates Judaism with irrational and Christianity with rational numbers, because Judaism's theological relationship to time cannot be expressed through reference to other moments of time, whereas Christianity's can in reference to Jesus.[53] Of interests here, however, is how the passage depicts the irrational numbers as equally necessary to the continuum, completing—indeed, "redeeming"—the rational numbers. Expanding on the example at the beginning of this section, mathematics provided Rosenzweig with an argumentative analogy to show the necessity of the irrational numbers of Judaism; they not only can determine the line via differentiation but also fill in the continuum to make it an "entirety," a "self-assured reality." This combination of rational and irrational numbers did not mean for Rosenzweig that Christianity and Judaism realize the infinite and achieve redemption within history. Instead, as is the case with *The Star of Redemption*, mathematics provides the terms to articulate that Judaism and Christianity can both represent the infinitude of redemption and work together, in their worldly acts, toward it as an ultimate, historically unreachable goal.

Taken together, knowledge focused on the world of experience and a more inclusive concept of time form the backbone of Rosenzweig's messianism, in which the activity of the individual in the present works toward a future messianic emancipation. For many of Rosenzweig's peers, including forerunners and early members of what became the Frankfurt School, such speculation about the links to a primordial past and a messianic future seemed dubious at best. "Rosenzweig babbles about God and the Creation of the world as if he had been there for it all," Siegfried Kracauer writes to Leo Löwenthal, "I strictly reject statements about the beginning of the world, the end of the world, etc."[54] Rosenzweig's emergent notion of redemption, however speculative, also elides the theological contribution of Islam, an elision systematized in *The Star of Redemption*.[55] For my concerns, however, Rosenzweig's critical contribution lies less in his theological claims than what the metaphorics afforded by infinitesimal calculus tell us about the possibilities of messianism: they refocus the light of messianism and the project of knowledge away from a passive expectance of divine intervention and onto our work in this world, the world in which we live. The Newtonian worldview was emancipatory, because it did not assume a singular, idealized and seemingly complete perspective of the world, but rather started the project of knowledge with the particularity and actuality of motion as it exists and as we as subjects, as finite beings, experience it. For Rosenzweig, a change in perspective of this nature was critical, in that it liberated the Jewish outlook on redemption from irrelevance, restructuring the concept of redemption not only to include but also to necessitate the Jewish contribution. It was also critical in the sense that it set the terms for a theory of society and, indeed, of knowledge that hinges on the reasonable actions and thoughts of the subject.

From Mathematics to Liturgy: Representation and The Star of Redemption

In this section, I turn to the transference of the metaphorics of subjectivity and motion in Rosenzweig's writing on mathematics into a messianism that takes the active engagement of the subject in the present as anticipatory of messianic emancipation. Even though mathematics plays a distinct role in addressing philosophical and theological matters in Rosenzweig's early work, the conceptual transference of mathematics into a messianism and, later, a messianic epistemology began in his major work, *The Star of Redemption*. Moving through analyses from creation to redemption, the book argues that the liturgical cycles of Judaism and Christianity anticipate the establishment of the Kingdom of God on earth, offering the faithful a glimpse of the divine "countenance" (*Antlitz*). The cyclical observance of religious holidays, the unending task of Christianity's mission in the world, and Judaism's eternal presence—in short, the actions of individual believers—represent in the finite world the eternity of the messianic age. *The Star of Redemption* entered into and shaped public intellectual discourse in the Weimar Republic and was read by and discussed among forerunners of the Frankfurt School, such as Kracauer, Scholem, Benjamin, and Adorno.[56] Indeed, it helped introduce a messianic element into theoretical approaches to culture, art, and history that flourished between the wars and formed the context out of which Horkheimer and Adorno's critical theory emerged.[57] These cultural and aesthetic theories viewed the present moment as a potential bridge to the messianic future. What the text adds to this messianic element in cultural and aesthetic theory is the idea that redemption and emancipation need not stand isolated from human activity. Instead, the establishment of a redeemed and emancipated world depends—as the metaphorics of subjectivity and motion drawn from infinitesimal calculus suggest—on the active participation of thinking subjects, even those who may seem to stand on the margins of so-called world history.

The messianic construction in *The Star of Redemption* borrows, as Leora Batnitzky has shown, on Hermann Cohen's concept of representation, which formed a key element of Cohen's work on both Judaism and mathematics.[58] According to Batnitzky, where Cohen often spoke of representation in terms of the German word *vorstellen* ("to imagine," or "to put before oneself"), Rosenzweig employs *vertreten* ("to step in for") "in a political and ethical sense."[59] While this shift already invokes the motion of the representative object (that "steps in," instead of being "put before one"), what attracted Rosenzweig to Cohen was also how Cohen used mathematics to illuminate a system of thought located in individual cognition—ideas similar to those discussed in "*Volksschule* and *Reichsschule*" and "*Urzelle*."[60] Indeed, the notion that the genesis of thought takes place in the mind of the individual underpinned the version of Neo-Kantianism that Cohen formulated in *The Logic of Pure Knowledge*. Here Cohen returned to Kant's theory of knowledge, but sought to displace the origin of knowledge from "experience" to a type of thought that has "no origin outside of itself."[61] For Cohen, in order for thought to be "pure," the thinking subject must generate—he uses the term *erzeugen*, which draws on the genealogical notion of procreation (*zeugen*)— the contents of thought within the confines of the mind and without recourse to space and time as given in the pure forms of intuition. The "creative sovereignty of thought" lay in what Cohen called "the logic or origin" and "the judgment of origin," which drew on the notion of the infinite judgment.[62] Accordingly, we judge "the something" (*das Etwas*) to be all that is not "the nothing" (*das Nichts*); hence, in this "something" we have a "pure" thought, without recourse to experience. For Cohen, it was understandable that this pure genesis of thought may remain obscure in the mind of his readers, so works such as *The Logic of Pure Knowledge* turn to mathematics as an example.

Cohen's idea that there is a representative relationship between mathematics and thought is significant for Rosenzweig's contribution to negative mathematics, because it will help link the activity of the individual with the messianic age. For Cohen, "infinitesimal-analysis is the legitimate instrument of the mathematical natural sciences": as *The Logic of Pure Knowledge* claims, "this mathematical generation [*Erzeugung*] of motion and, hence, nature is the triumph of pure thought."[63] What Cohen means here is that the new calculi developed by Leibniz and Newton offer not an ontology of the physical world outside the mind as infinitely small particles, but rather an "instrument" to demonstrate how ideas can be "generated" inside it: "Leibniz designates the new concept [of the infinitely small] with dx. This dx, however, is the *origin* of x, with which analysis calculates, and which is the representative of the finite. Therefore, this definition even seizes upon the judgment of origin in order to define the infinitely small. And so it is with the infinitesimal, as with the fluxion, the great *example* of the fundamental meaning of the judgment of origin."[64] For Cohen, how mathematics gener-

ates "finite" quantities out of the "infinitely small" offers an analog for—
even an example of—the genesis of pure thought (the "judgment of origin")
in his philosophical system. In the passage, Leibniz's "differential" (*dx*, similar to Newton's fluxion) serves as the "the *origin* of x" in the process of
differentiation because these infinitely small quantities "generate" finite
quantities—or, as in the case of Newton, the point "generates" the line
through continuous motion. The passage also defines the relationship between
this process and pure thought as one of representation. The infinitesimal is
the "great example" of the "judgment of origin," the nonfinite "stands in
for" (*vertreten*) the "ground of the finite"; hence, the "triumph of pure
thought," in Cohen's words, can be demonstrated "on the paradigmatic example [*Musterbeispiel*] of infinitesimal calculus."[65] In essence, the creation
of the curve from infinitely small increments of motion in differentiation
illustrates and renders legible the genesis of pure thought. Indeed, Cohen's
turn to infinitesimal calculus—and not a more contemporary branch of
mathematical thinking, such as set theory—is due precisely to the former's
ability to serve as an example of a generally understood methodology essential to the Newtonian natural sciences. This process was or, at least,
should have been intelligible to all those, as Rosenzweig proposed, educated
at the *Reichsschule*.[66] For Cohen, mathematics' utility lay in representing, in
exemplifying what otherwise may be too obscure for thought to grasp—
such as, its own origin.

In order to establish a connection between the individual and the messianic, *The Star of Redemption* combines Cohen's idea of representation with
the metaphorics of motion and subjectivity drawn from mathematics. The
work begins with subjectivity, "the fear of death," as the insoluble and in-
delible problem that sheds doubt on idealism's "blue haze of the thought of
the All" (R 2:3). Yet the book does not abandon the philosophical project of
knowledge, but rather seeks to know "the All," as Pollock puts it, from "a
quintessentially human standpoint" instead of the "Absolute standpoint of
the Idealists" (4).[67] The starting point of knowledge thus must lie within the
philosopher. The text finds this point of departure in three hypothetical aspects of knowledge—God, World, and Self—about which we know "nothing whatsoever" (23). As in "*Volksschule* and *Reichsschule*" and, now, with
Cohen's help, mathematics serves as the "guide" out of this nothingness:

> It was Hermann Cohen . . . who first discovered in mathematics an *organon* of thought, precisely because it generates [*erzeugt*] its elements not out of the empty nothing of the one and general null, but out of the particular nothing of the differential, each assigned to the very searched-for element. The differential combines in itself the properties of the nothing and the something; it is a nothing that hints at a something, its something, and, at the same time, a something that slumbers in the womb of the nothing. In one, it is a quantity as it flows into that without quantity and has, in turn, all the properties of a finite quantity on loan from that which is "infinitely small," with one exception: this itself. It thus draws its power to create reality, first from the violent negation, with which it breaks the womb of the nothing and then equally from the calm affirmation of all that borders on the nothing, to which it remains attached infinitely small, as it is. The differential thus opens up [*erschließt*] two paths from the nothing to the something, the path of the affirmation of that, which is not nothing, and the path of the negation of the nothing. (23)

Striking about this passage is how motion and subjectivity inform representation on both the rhetorical and conceptual level. For instance, in the first sentence, the passage codes "mathematics" along the lines of subjectivity: it is an "organon" for the thinking subject that "opens up" (as in "makes accessible") for thought two ways of grasping the "generation" of the something from the nothing. The "differential," the ratio of instantaneous vertical to horizontal movement, signifies infinitely small quantities as it—as in the metaphorics of motion—"flows" from quantity into "that without quantity" and "combines" the properties of existence and nothingness. As in Cohen's use of mathematics to illustrate pure thought, the differential is the "nothing" that "hints" at its opposite, having "on loan" these properties "from that which is infinitely small." In this passage, mathematics dynamically combines both the nothing and the something, the nonfinite and the finite "in one": the differential presents itself as a something to the subject and, at the same time, stands in for its limit point, the nothing.

Furthermore, this passage reveals the specific type of negativity that mathematics helps Rosenzweig address through the metaphorics of subjectivity and motion. In the terms laid out by *The Star of Redemption*, the differential articulates the seemingly paradoxical somethingness of nothingness—for death and the hypothetical aspects of knowledge of God, World, and Self.

This nothingness, however, is not "the empty, general nothing" (R 2:22); such nothingness would be more akin to Hegel's notion of the nothing as pure absence of determination, an absolute nothing.[68] Likewise, the nothingness in the text is not the absolute Platonic divide between mathematics and human mind and language (see chapter 2). Instead, as in the previous excerpt, the differential emerges out of the void of "the particular nothing," because the "particular nothing" is particular to something else and, thus, contains a reference to a something—in the case of the differential, an infinitesimal quantity that references a finite quantity (23). Analogously, creating knowledge of the three nothings of God, World, and Self make up the first epistemic steps that unfold in *The Star of Redemption*. Mathematics helps deal with this negativity on two fronts: it shows how one can generate positive knowledge out of specific nothingness and represents this knowledge when other languages cannot.

The ways in which mathematics addresses negativity shapes, as I explore in the remainder of this chapter, the theories of language and liturgy proposed in *The Star of Redemption*, providing the basis for Rosenzweig's messianism and messianic epistemology. Consider first Rosenzweig's theory of language, which originates in the description of the three hypothetical elements (God, World, and the Self) "searched-for" in part 1 of the book. For Rosenzweig, any knowledge that we have of these elements that exist before creation stems from infinitesimal calculus—namely, the two "paths" that the differential "opens up." For instance, the first attribute of God derives from the "path of affirmation": "the affirmation of the not-nothing [*des Nichtnichts*] thus circumscribes as an inner boundary the infinity of that which is not nothing. An infinite is affirmed: God's infinite being" (R 2:29). As the "affirmation of that, which is not nothing," the differential models how we can know God's being to be everything with the exception of nothingness itself, which is thus the infinite totality of what is. In *The Star of Redemption*, "the power of the Yes" that affirms God's infinite being represents "the primordial word [*Urwort*] of language, one of those primordial words through which—not even sentences, but rather the very words that form sentences, the word as part of the sentence, become possible" (29).[69] Like the "primordial phenomenon" of the natural world in "*Volksschule* and *Reichsschule*," the "primordial worlds" ("Yes," "No," and "And") set the conditions for the possibility of language. As a conceptual "tool," the differential "opens up" to the thinking subject the otherwise unintelligible "primordial word" (*Urwort*), not as a divine gift, nor as raw data of the world, but rather as a product of the thinking subject itself.

For Rosenzweig, mathematics' ability to derive the "Yes" as a hypothesized "primordial word" is not limited to God's being alone, but rather underpins the schematic substructure of language as a whole. Out of this positing of the "Yes" as the first "primordial word," the text depicts how the other two "primordial words" unfold systematically from the differential: first, from the "No" revealed by the "path of the negation of the nothing" that represents "God's freedom," and, second, from the "And" that signifies the sequential combination of "Yes" and "No" and represents "God's vitality."[70] As was the case with Cohen, the rhetorical focus on mathematics' ability to represent expresses itself, according to Rosenzweig, not only conceptually but also symbolically. In all three books of part 1, the text derives and depicts the relationship of affirmation and negation in "familiar logical-mathematical symbols," in particular in "the algebraic letters and the equals sign": "We symbolize the total lack of relation with the simple x or y; the relation of the subject to a predicate with $y =$, thus the definition as it regards an equation yet to be assigned; the equation with $= x$ as it regards a definition yet to come." Using, as in "*Urzelle*," "A" to represent generality and "B" to represent particularity, *The Star of Redemption* systematically generates the properties of God, World, and the Self (table 3.1). Here mathematics offers the subject a language, first the differential and then algebra, to express the natures of God, World, and Self that remain inexpressible in other languages. As charted out by the metaphorics of subjectivity, mathematics is a symbolic tool for thought to depict the otherwise unknowable origin of knowledge in the thinking subject and to the thinking subject.

As *The Star of Redemption* moves from the prelinguistic conditions afforded by mathematics to language itself, the metaphorics of subjectivity and motion become intellectually generative in Rosenzweig's thought. This transition constitutes a key dimension of Rosenzweig's contribution to negative mathematics, as the dynamic ability of the differential to represent informs the linguistic relationship among God, World, and Self and, as I

TABLE 3.1. The Properties of God, World, and Self

Element	Primordial Word	Attribute	Symbol
God	Yes	Divine being	A
God	No	Divine freedom	A=
God	And	God's vitality	A=A
World	Yes	Worldly order	=A
World	No	Worldly plentitude	B
World	And	Reality of the World	B=A
Self	Yes	The Self's particularity	B
Self	No	The Self's will	=B
Self	And	The Self's independence	B=B

will show shortly, the liturgical relationship between the present and messianic future. The transition from a mathematical pre-language to a language used by humans in the world also circumscribes the limits to mathematics as an intellectual instrument. "In the self-expressing, and self-revealing world, another medium takes over the assignment of depicting meaning, the role of the organon to provide symbols," the text claims, "a science of living sounds must take the place of a science of silent symbols [*an Stelle einer Wissenschaft stummer Zeichen muß eine Wissenschaft lebendiger Laute treten*], the teaching of the forms of words, grammar, takes the place of a mathematical science" (R 2:139). The "forms of words" and "grammar" that follows from them provides the central thesis that the text unfolds in book 2, linking revelation to speech. But even as "the mathematical language of symbols, in which we could depict the emergence of the elements, fails here," the metaphorics surrounding the infinitesimal calculus remain—at least in the structure of Rosenzweig's argument—in effect (138). If, for example, grammar "takes [the] place" (*an Stelle treten*) of mathematics, then the "silent symbols" of mathematics "as a science" stand in (as in *vertreten*) before the "living sounds" of grammar take over, similar to how, for Cohen, the infinitesimal first "stands in" for the finite (139). In *The Star of Redemption*, mathematics fulfills the charge of exemplarity and utility that Rosenzweig first found in infinitesimal calculus in "*Volksschule* and *Reichsschule*," but its methods and tools remain within a circumscribed boundary. Instead of continuing the application of mathematics and mathematizing language, the text draws on the intellectual conditions that they rendered legible to theorize language as focused on the subject and the messianic future symbolized by its actions in the world.

In particular, the metaphors of motion help shape Rosenzweig's delineation of the "living" language in which God creates the World, God reveals Himself to the Self, and the Self relates to the World via interpersonal love. Similar to the process by which the differential "generates" the two paths leading to the "primordial words," these "primordial words" likewise generate the "root words" of language: "Out of the mute primordial words that merely accompanied thought made visible in the algebraic symbols," Rosenzweig explains, "must spring forth [*entspringen*] audible words, in a certain sense root words [*Stammworte*]" (R 2:140). In this passage, the emergence of "audible" language invokes the dynamism of the mathematical idea of generation and origins, as the verb *entspringen* shares the root verb *springen* with *Ursprung*. As in the discussion of God, World, and Self in part 1 of *The Star of Redemption*, part 2 systematically derives "root words" (*Stammworte*, and one "root sentence") out of these "primordial words." According to Rosenzweig, the move from "mute" mathematical pre-language to "audible" language occurs through the analysis of Genesis, which produces the first "root word" that articulates the relationship between God and World. God's "positive evaluation" of creation—God's six-fold exclamation "Good!" in Genesis 1—is "nothing more than the primordial Yes become audible" (141–142). Extrapolating on this logic, the text then shows how the "root words" constitute three grammatical moods, which correspond to the theological categories of creation, revelation, and redemption (table 3.2).[71] In these grammatical modes, God affirms creation, God reveals Himself to the individual in the act of divine love, and the Self works to redeem the World through neighborly love. Absent in this depiction are the "flowing quantities" of the differential and the "mute" signs of algebra. Instead, these metaphorics operate on the level of Rosenzweig's argument itself, as his book charts the dynamic genesis and motion of a system of thought out of and around the individual.

TABLE 3.2. Rosenzweig's Grammar, Derived from the Primordial Words

Category	Mood	Primordial Word	Root Word	Example
Creation	Indicative	Primordial yes	Adjective	Genesis
Revelation	Imperative	Primordial no	Self	Song of Songs
Redemption	Cohortative	And	Root sentence	Psalm 115

This point of transition between mathematics and language in *The Star of Redemption* already opens up avenues for addressing instrumentality and the position of the subject in theories of culture and language. For Rosenzweig, mathematics was a powerful instrument, but it also had its limits. In *The Star of Redemption*, we have a theory of language that borrows from mathematics to formulate the conditions for language's existence, but it does not turn language into mathematics itself. Instead, Rosenzweig's thinking actualizes in language the cognitive options that mathematics first renders legible: the affirmation, negation, and conjunction that articulate the attributes of God, World, and Self. Furthermore, the vision of language produced in *The Star of Redemption* contains, in the words of Eric Santner, "an openness to the alterity, the uncanny strangeness, of the Other": the expression of neighborly love between the Self and Others in the world reflects, for Rosenzweig, the divine love revealed by God to the Self.[72] Embedding the relationship between God and Self in the interactions of the Self with the World, Rosenzweig's theory of language locates the emancipatory work of language in ongoing human interaction with Others in the world. This possibility not only reflects the metaphorics of motion and subjectivity, but also suggests the idea that Rosenzweig's messianism bears out in full: the action of the individual serves as the agent of redemption in the everyday world.

As the theory of language in *The Star of Redemption* emerges from the possibilities opened up by mathematics, Rosenzweig's theory of liturgy and redemption continues to draw on the metaphorics of motion and subjectivity in formulating a messianism based on dynamic subjectivity. Part 3 of *The Star of Redemption* demarcates that the possibility of working toward the realization of the Kingdom of God on earth is fulfilled by neither "the silent keys" of mathematics (proper to the primordial world) nor the "revealed symbols" of grammar (proper to the lived world), but rather the liturgical cycle and "the prayer for the coming of the Kingdom" (R 2:326–327). The text thus instantiates metaphors of motion in its progression from mathematics to language to liturgy. It also distinguishes the type of knowledge proper to liturgy in contrast to the representative relationships proposed by language and mathematics:

> The forms of liturgy, however, do not possess this simultaneity [of language] in that which is to be known though them; indeed, they anticipate [*nehmen vorweg*]; it is a future that they make into a today. Thus, they are neither keys nor mouth of their world, but rather representatives [*Vertreter*]. They represent to knowledge the redeemed supra-world; knowledge knows only them; it does not see beyond them; the eternal hides behind them. They are the light in which we see the light, silent anticipation of a world illuminant in the silence of the future. (327)

This passage is unique because it encapsulates the central idea in Rosenzweig's contribution to negative mathematics. Human work in the present, "the everyday, every-week, and every-year repetition" of the "forms of liturgy," prayer in particular, "anticipates" and serves as a "representative" for the Kingdom of God (325). By participating in the unending "repetition" of liturgy, my actions mirror the eternity of the "redeemed supra-world." This idea, however, realizes as a theory of redemption the metaphorics of subjectivity and motion that we have seen in association with infinitesimal calculus: while the "keys" (*Schlüssel*) of mathematics "opened up" (*erschließen*) knowledge of the pre-world to the knowing subject, now the action of the subject, my participation in the liturgical cycles, "represents to knowledge," stands in (as in the term *vertreten*) in the present for future redemption. For Rosenzweig, liturgy was the "differential" of theology, just in reverse. Where the differential represented finitude and nothingness at the same time, liturgy "anticipates" and "represents" in finite experience the infinitude of redemption, of the experience of God. This action in the present that stands in for future redemption is Rosenzweig's messianism, but it is also the product of negative mathematics: in the same way mathematics turned the indecipherable nothing into knowledge, knowledge anticipates a redemption through human action in liturgy and prayer, behind which "the eternal hides." Indeed, as "silence" (*Stille* and *Schweigen*), liturgy even returns us to the "silence" (*Stummheit*) of mathematics.

Before turning to the more general implications of Rosenzweig's messianism, I want to address two concluding points in the text's discussion of Judaism that illuminate the critical contribution of negative mathematics in Rosenzweig's thought. In creating a system of philosophy centered on the subject, *The Star of Redemption* puts into practice the metaphorics of subjectivity that we first saw in *"Volksschule* and *Reichsschule."* Accordingly, "the grasping" of the messianic, redeemed world through the subject's participation in the liturgical cycles "takes place in the illumination of prayer. We saw how the path rounded itself off here into the yearly cycle and, by this means, how the All, when just this closure is prayed for, offers itself immediately for the beholding" (R 2:435). This is not to say that religious practice realizes the teleological end of history in the present. Instead, through anticipation and representation, the individual enters in the here and now into a relationship with transcendence, "beholds" the All through the "rounded off" repetition of the "yearly cycles" of liturgy—"closure" hoped for through prayer. Significant here is how the text specifies that this subjectivity includes Jewish subjects, who, along with Christians, "share" in the "whole truth" of the "All" (437). In Rosenzweig's words, "the fire or the eternal life" of Judaism and its liturgical cycle anticipates eternity by cycling through the "beginning, middle, and end of this national people" in "every new generation, no with every new yearly cycle and with every new year" (352). For Rosenzweig, Judaism may stand on the margins of world history since the destruction of the Second Temple, having ceded, as in the 1913 conversation that prompted him to consider conversion, worldly power to Rome and Christianity. Judaism's contribution to redemption, however, lies not on the stage of history in *The Star of Redemption*, but rather in representing the completion of the theological categories through its liturgical cycle of annual festivals: creation with the Sabbath, revelation with Passover, Shavuot, and Succoth, and redemption with Yom Kippur. A mirror image of the differential that relates the nothing to the finite something in infinitesimal calculus, it is Jewish liturgy that, despite its seeming historical marginality, brings into experience the eternity of the messianic age to the extent that it can be known and experienced by the individual as a finite, creaturely being.

For Rosenzweig's messianism, this completion of the Jewish liturgical cycle demonstrates the necessity of Judaism. As in the two examples that compared Judaism and Christianity as points on a curve and irrational and rational numbers, *The Star of Redemption* frames the necessity of Judaism in contrast to Christianity, "the rays or the eternal path," in mathematical terms. Christianity anticipates the eternity of redemption by spreading "the evangel" ("good news") in the world between the first and second coming of the Messiah (R 2:405). The expansion of Christianity thus stands in for eternity as open-ended, represented in Christian liturgy by the holidays of creation (Sunday) and revelation (Christmas, Easter, and Pentecost), but lack of a holiday of redemption. The continued existence of Judaism remains necessary, in Rosenzweig's eyes, as a reminder of the incompleteness and, hence, infinity of Christianity's expansionary task: "this being-there [*Dasein*] of the Jew forces the thought upon Christianity at all times, that it has arrived neither at the goal nor at the truth, but rather remains—on the way" (459). To illustrate the contribution of Judaism's "eternal life" to keep Christianity on the "eternal way," the text turns to the implications of the metaphorics of motion:

> [The Christian and Jewish relationship to eternity] is as different as the infinity of a point and of a line. The infinity of a point can only consist in that it is never wiped away. It thus sustains itself in the eternal self-preservation of propagating blood. The infinity of a line, however, ceases to be, when it is no longer possible to continue expanding. It consists in the possibility of uninhibited expansion. Christianity as eternal path must always expand further. The simple preservation of its contents would mean abandoning its infinity and, hence, death. Christianity must missionize. . . . It propagates itself, in that it expands itself. (379)

With a continuum as continuous motion, there is more than one relationship to the infinite: the intensive perpetuity of Judaism's point and the extensive endlessness of Christianity's line. Indeed, if we follow the logic of the metaphor, the passage suggests that Judaism's points are constitutive of Christianity's line; as Cohen put it: "The point is no longer only the end, but rather much more the beginning of the line."[73] In Rosenzweig's messianism, the infinity of the point and the line reveal and embody the multiplicity of our relationship to redemption. What is generative about negative mathematics in the text is the way it illuminates more capacious philosophical and theological concepts—ones that not only accommodate but also necessitate different perspectives to bring into human experience that which, ultimately, stands beyond it.

In *The Star of Redemption*, the metaphorics of subjectivity and motion that Rosenzweig associated with mathematics transform into a messianism that depends on the active engagement of the individual. The dynamic capacity of the differential to represent the nothing as a something allowed Rosenzweig to remove, as Scholem later wrote, "the apocalyptic thorn from the organism of Judaism," bringing redemption into the here and now of religious experience while pushing its realization out of historical experience.[74] This messianism is the contribution of Rosenzweig's negative mathematics as a theory of culture that takes our work toward emancipation as representative of an unrealizable redemption. Even as many of the later members of the Frankfurt School may have dismissed Rosenzweig's reorientation of philosophy around the subject, his messianism reemphasizes the role of the individual critic in, as Walter Benjamin describes it, the "weak messianic power" operative in cultural criticism.[75] Where philosophy no longer considered the individual and theology obscured pluralistic claims to redemption, mathematics gave voice for Rosenzweig to an individual, a Jewish contribution to salvation, revealing that even those groups that stand on the sidelines of world history still play a role in redeeming the world. For cultural criticism today, this messianism indicates that the possibility for transcendence, the potential of emancipation, depends not on messianic interruption, but the actions of individuals in the here and now. Indeed, by making redemption a product of individual action, the text suggests that the work of the subject—to understand and to work toward the goal of emancipation—occupies a central place alongside mathematics in the project of knowledge.

Rosenzweig's Messianic Theory of Knowledge and Critical Theory

This chapter has shown, largely in theological terms, the ways in which mathematical techniques for dealing with negativity helped Rosenzweig forge a messianism of the present. And yet Rosenzweig's messianism was also always a question of knowledge, in the terms laid out by *The Star of Redemption*, of "knowing the All" (R 2:3). Whereas Horkheimer and Adorno saw in mathematics the reification of knowledge into an instrument of control and domination, Rosenzweig found in mathematics a tool to argue for the epistemological significance of individual action and belief. This epistemological dimension of Rosenzweig's thought becomes clearer in his essay "The New Thinking" ("Das neue Denken"), published in 1925 as a companion piece to *The Star of Redemption*. The text of "The New Thinking" not only specifies that the text should be read less as a "Jewish book" than "just a system of philosophy" but also expands the contribution of negative mathematics in Rosenzweig's messianism into a "messianic theory of knowledge" (*eine messianische Erkenntnistheorie*; R 3:139–140 and 159). In terms of the project of negative mathematics, Rosenzweig's messianic theory of knowledge picks up where Scholem's negative aesthetics left off. Scholem's negative aesthetics took the restriction of language from mathematical logic as a formal strategy for poetry, indicating the possibility for historical continuity despite catastrophe. For Rosenzweig, the representational link between finitude and infinitude forged by infinitesimal calculus signified the contribution of the work and beliefs of individuals not only to redemption but also to epistemology. As with the uncreated world, mathematics lends a language to a more expansive theory of knowledge, to which I turn in these closing remarks. Rosenzweig's messianic epistemology redefines knowledge to include the knowledge transmitted by cultural traditions and produced by cultural theory—forms of knowledge, in other words, that cannot be "proved" in the mathematical sense of the word.

At its core, negative mathematics in Rosenzweig's thinking is a politics of epistemology. A similar politics was at work for Scholem in what negative mathematics revealed about language: language without representation can signify events and peoples that themselves evade or lack poetic and historical representation. For Rosenzweig, the dual sense of representation offered by infinitesimal calculus opens up the concept of redemption to include a Jewish contribution and, in "The New Thinking," becomes a more inclusive epistemology:

> Thus truth ceases to be what "is" true and becomes that which—wants to be *verified* as true. The concept of the verification of truth becomes the basic concept of this new epistemology, which takes the place of the old epistemology's noncontradiction[-theory] and object-theory, and introduces, instead of the old static concept of objectivity, a dynamic concept. The hopelessly static truths like those of mathematics, which the old theory of knowledge took as a point of departure without ever really going beyond this point of departure, are to be understood from this perspective as the—lower—limit case, as rest is understood as the limit case of motion, while the higher and highest truths can be grasped only from this perspective as truths, instead of having to be relabeled into fictions, postulates, or needs. From those unimportant truths of the stripe "two times two is four," on which people easily agree with no other expense than a bit of brainpower—for the multiplication tables a bit less, for the theory of relativity a bit more—the path leads over the truths, for which a person is willing to pay, over to those, which a person cannot verify in any other way than with the sacrifice of his life, and finally to those, whose truth only the commitment of the lives of all generations [*der Lebenseinsatz aller Geschlechter*] can verify. (R 3:158–159)

Here we recognize the metaphorics of motion and space at work: the passage proposes a "dynamic" epistemology in which truth depends on me, the subject, verifying it as true.[76] The epistemological imperative of this passage reads that we must no longer consider as knowledge (*Erkenntnis*) only forms of knowledge that adhere to mathematical or natural-scientific modes of knowing. We must also take account of the knowledge that I as a subject and we as groups make true through action and belief, but whose ultimate truth becomes visible only at the unreachable end of "all generations." What Rosenzweig's negative mathematics offers to contemporary debates about the humanities is the idea that we should not dismiss mathematics and theoretical physics as valid ways to produce knowledge about the world, but rather uphold the epistemic contribution of action, belief, and critique by reframing the debate over what counts as knowledge.

The idea in negative mathematics that human action in the world anticipates redemption in Rosenzweig's messianism allows us to flesh out how a new theory of knowledge could view and evaluate these actions in epistemological terms. As the previous quote from "The New Thinking" suggests, this theory would view knowledge not as what is absolutely proved, but rather what individuals and groups "verify" in and through their experience. For Rosenzweig, verification did not simply mean that any idea "verified" in experience automatically counted as knowledge, nor did it imply that theoretical statements only became meaningful, as Carnap later argued, when "verified" by experience.[77] Instead, the "truth must be veri-fied [*be-währt*]," *The Star of Redemption* claims, "in that one lets the 'whole' truth rest on itself and, nonetheless, recognizes the portion, on which one holds themselves, as the eternal truth" (R 2:437). Current usage of the verb "to verify" tends to obscure the meaning of the Latin roots *vērus* ("true") and *facere* ("to do, to make") and the German *bewähren*, which also means "to make true" as the prefix *be-* implies. I "make" theological truth "true" by participating in and, thus, confirming the "eternal truth" of the liturgical cycles, which brings into the here and now my "portion" of the "'whole' truth" that remains inaccessible to me as a finite being. Like a theory of history that would include voices that lack historical representation, knowledge as verification would include critical and marginal voices that stand in perhaps more oblique relationships to history or mainstream forms of knowledge production such as, in the contemporary epistemological climate, the natural sciences. Taking from mathematics a sense of the efficacy of representation, this messianic theory of knowledge accommodates as knowledge my work toward an emancipatory cause in this world based on its relationship to the absolute goal of emancipation that, nonetheless, cannot be said to stand in a historical relationship with the present.

Such a messianic theory of knowledge draws on the mathematical approach to negativity in as much as verification displaces the criterion of what counts as knowledge beyond human experience. As Rosenzweig explains in a revealing albeit jargon-filled missive to Kracauer:

There is a way whereby a time, an -ism, or something similar can become absolute [*eine Zeit, ein Tum oder dergleichen absolut werden kann*]. But, it eludes knowing by proof or demonstration . . . knowledge; for its object is neither rational nor "irrational," but rather much simpler: not yet there at all. It's a question of a becoming-absolute [*Absolutwerden*], not of a being-absolute [*Absolutsein*], a question of the one by the grace of the other; in a certain sense, a being-absolute by partial payments [*ein Absolutsein auf Abschlagzahlung*]. It depends on whether the installments are paid on time—historically or, respectively, personally on time. Hence, logically (new-logically) speaking: it depends not on proof, but on verification [*nicht auf den Beweis, sondern auf die Bewährung*].[78]

In this, as in the previous excerpt, messianic epistemology expands on forms of knowledge that we "prove" or "demonstrate"—the signature of logical, mathematical, and philosophical credibility—but this expansion does not mean its contents are "irrational."[79] Instead, it functions as a piece-by-piece accumulation of knowledge, "partial payments." Such "payments" are the work that individuals and groups dedicate in the real world toward the realization of an idea that cannot enter into historical experience because it is "absolute." So conceived, knowledge would include what we "prove" to be "absolute" (such as $2 + 2 = 4$), but also processes of "becoming" absolute, such as the ongoing work of emancipation. Indeed, Rosenzweig worked to put the political imperative of his messianic epistemology into practice. He dedicated his own subjective action in the world toward rebuilding Jewish education in the form of the Freies Jüdisches Lehrhaus and argued for the necessity and relevance of Judaism in the scheme of redemption.[80] As neither the political power of religious beliefs nor the fight against oppression and the work of social justice show signs of leaving our contemporary cultural horizon, further articulating the epistemological contribution of these endeavors seems as pressing a task as any.

What remains significant about the development of Rosenzweig's messianism and messianic epistemology is that it did not reject mathematics outright, declaring it somehow at odds with or threatening to philosophical, theological, or cultural thought. Rather, mathematics and the concepts of infinitesimal calculus illuminated and represented alternative perspectives on concepts such as subjectivity, time, and redemption that were (and still are) central to critical theory yet evaded the languages of philosophy and theology. Rosenzweig's thought thus offers an example of how cultural criticism can borrow from mathematics to illuminate its concepts without mathematizing culture or criticism itself. As infinitesimal calculus digs into and reveals the multiplicities of subjectivity, motion, and representation, the theoretical work done by mathematics in Rosenzweig's thought begs the question of what other mathematical tools—both in the early twentieth century and today—may be available for and useful in cultural criticism. These tools could help theorists think through concepts that remain obscure in aesthetic and cultural theory, as fractal geometry illuminates the theory of the novel for Wai Chee Dimock.[81] Can mathematics help us construct more capacious versions of these concepts as well? Do conceptual tools exist that allow us to intervene more immediately in a project of emancipation, in the service of which theories of culture and art work? Notwithstanding Rosenzweig's dismissal of the field, one such possibility, which I will explore in chapter 4, lies in geometry and the aesthetic dimension of cultural critique itself.

FOUR

Geometry: Projection and Space in Siegfried Kracauer's Aesthetics of Theory

On January 11, 1920, Siegfried Kracauer revealed something that must have surprised his then friend and mentor, Margarete Susman: "More and more, my thinking is approaching higher mathematics."[1] Against the intellectual doubt and skepticism that plagued the postwar era, "higher mathematics" and, in particular, geometry offered Kracauer a framework to evaluate and compare the systems of thought that attempted to fill the void of meaning that war, revolution, and the collapse of an empire had left behind. However, for another of Kracauer's friends, Theodor W. Adorno, the equation of "thinking" and "higher mathematics" by the logical positivists meant, just over a decade later, the expulsion of language from philosophy and threatened a wholesale "liquidation of philosophy" itself.[2] The idea of excluding language from philosophy was deeply troubling for Adorno and Horkheimer, because language's contribution to philosophy—philosophical style—constituted an irreducible element of thought, often reflected in the tortuous prose that describes their version of critical theory (see chapter 1). And yet a new philosophical style was exactly what geometry, the mathematical study of space, presented to Kracauer in the early 1920s. For Kracauer, the ways that different branches of geometry, such as Euclidian geometry, combined a logically rigorous study of space with a sense for the concrete materiality of space offered a novel approach to negativity—namely, the divide increasingly separating the experience of modernity from the available cognitive tools to grasp modern existence. As a metaphorics of space and method of projection, geometry, I contend, transformed in Kracauer's Weimar-era writings into an aesthetic program that took the space of the text, the composition of theory itself, as a means to intervene in cultural debates. The material composition of thought—the aesthetics of theory—held the potential, at least for Kracauer, to realize the promise of the Enlightenment and create a society based on reason.

This chapter explores the development of a metaphorics of space and interpretive method based on geometric projection in Kracauer's writing and their implications for his vision of cultural critique and critical theory as a whole. Kracauer's reputation in critical circles has often rested on his pioneering work as a film theorist, his studies on propaganda, and his friendship with and erstwhile mentorship of Adorno.[3] Here I turn to the critical potential of his earlier work as a sociologist and feuilletonist for the *Frankfurter Zeitung*, in which he reveals himself as a keen cultural observer and critic of modernity in the early Weimar Republic. These early texts developed what is known as Kracauer's method of cultural critique, which blended a criticism of material objects with a theory of their function in society.[4] For Kracauer, geometry's approach to negativity helped him think through an intellectual crisis, in which, at the advent of a new century, inherited modes of analysis such as academic philosophy and cultural practices such as religion no longer addressed modern life. Where the philosophy of mathematics indicated for Scholem a language lacking representation and infinitesimal calculus offered representational tools for Rosenzweig, geometry reconnected the material and the logical world for Kracauer, lest materiality abandon reason altogether or reason disappear into the obscurity of pure thought. In particular, a set of spatial metaphors drawn from geometry pointed Kracauer to the interstitial region between rule-bound logic and the contingency of experience. To be sure, scholars have noted such geometric motifs in Kracauer's writing.[5] Here, however, I examine the moment at which the interstitial area opened up by geometry translated for Kracauer into a method of cultural-critical projection that rendered legible the metaphysical meanings behind mass phenomena such as the Tiller Girls, detective novels, and the modern city. In these texts, how geometry bridged the abyss between metaphysics and materiality became a literary strategy in his writing that drew attention to the rational construction of the text and, thus, sought to promote rational thought in the mind of his readers.

Bridging the logical and material worlds, thought and being, was of the utmost importance for Kracauer, because reason and meaning were the aspects lacking in the cultural and physical expressions of modernity. For Kracauer, the modern world was, in terms he often borrowed from Georg Lukács, "abandoned by God," the age of "transcendental homelessness," in which neither Judaism nor Christianity offered Kracauer a viable point of intellectual orientation. For those acquainted with Weimar modernisms, this is a familiar narrative of modernity, in which, as Miriam Hansen writes, the present "appears as the endpoint of a process of disintegration, spiritual loss, and withdrawal of meaning from life, a disassociation of truth and existence."[6] The possibility of reconnecting this disjunction of meaning and existence was the generative negativity activated for Kracauer by the combination of logic and materiality in geometry. As a mathematical technique, geometric projection suggested ways to read the mass-produced and often ephemeral products of a society—for example, public spaces, films, and dance revues—as indicative of that society's place in the metaphysics of history. As a mode of deciphering these metaphysical meanings, geometry thus helped Kracauer construct a vision of history in which thought could intervene in process of Enlightenment and help establish a reasonable and inclusive society.[7] While modern culture seemed stuck within the "murky reason" (*getrübte Vernunft*) of capitalism, geometry's synthesis of thought and experience suggested literary techniques that could confront readers with the contemporary stagnation of reason and, through this confrontation, further the process of history.[8] For Kracauer, it was the job of the cultural critic, the societal observer—indeed, of a Jew on the margins of society—to reveal the meaning of the meaningless products of mass culture and advance the project of Enlightenment.

The idea that the aesthetics of critique could work toward reasonable society constitutes the contribution of Kracauer's negative mathematics to discussions of critical theory in the present. Although histories of critical theory have pointed to the significance of aesthetics and aesthetic mediation for the first generation of critical theorists, I believe Kracauer's use of geometry reminds us of the deeper dimensions of "aesthetics" as a critical term.[9] In the present context, the term *aesthetics* refers less to the pleasurable appearance of things or even the scientific study of beauty than to the idea of perception as implied by the Greek term *aisthesthai*.[10] Like geometry, aesthetics was for Kracauer the liminal point of contact and interaction between the material world and cognition. This understanding of aesthetics informed Kracauer's cultural critique, which was aesthetic in as much as it analyzed forms of culture and art as well as performed an analysis on the textual level of form—an idea later canonized by Adorno's *Aesthetic Theory* (*Ästhetische Theorie*, 1979).[11] Kracauer's negative mathematics shows how cultural critique can be not just an abstract theory of culture but also an aesthetic venture that seeks to change society through the materiality of the text. Aesthetic critique uses its composition and presentation to confront readers, on the level of form, with the problematic rationality of contemporary society, trigger critical reflection, and, thus, work toward a reasonable society. In the aesthetic dimension of Kracauer's writing, we again recognize the theological impulse of critical theory, which sees the cultural critique offered by those on the margins of mainstream society as working in the service of emancipation and redemption. Critique that is aesthetic can itself be emancipatory in as much as it includes the voices of critics who lead, in Kracauer's words, the "extra-territorial lives" of displacement, diaspora, and exile and perhaps, even those perspectives (such as Kracauer's) that have remained in the shadow of the Frankfurt School.[12]

In charting the emancipatory potential of critique that Kracauer found in negative mathematics, one comes across four meanings of the term *geometry* that interconnect as a metaphorics of space. Taken together, these meanings constitute geometry's approach to negativity that, for Kracauer, traversed the rift separating thought and experience. The first meaning of the term refers to geometry—and, in particular, Euclidean geometry—as the paradigm of a strict logical system (similar to how Scholem and Rosenzweig use the term). Accordingly, our knowledge of space unfolds out of self-evident axioms in such a manner that "every proposition constructed out of the designated axiomatic concepts" is, as Kracauer's interlocutor on the subject (Edmund Husserl) writes, "a pure formal implication of the axiom."[13] The second meaning of *geometry* refers to drafting and engineering techniques that Kracauer, trained as an architect, would have encountered and employed in his university courses on descriptive geometry (*darstellende Geometrie*).[14] One such method, projection, allows us to depict three-dimensional space as a two-dimensional drawing that is more amenable to neat mathematical transformations. For Kracauer, running projection in reverse offered an interpretive method that "projects" the material products of society into the metaphysical space of history in order to decipher their meanings. On the other side of the spectrum, the third meaning of the term is the idea of a "natural geometry," which René Descartes used to describe our innate ability to perceive direction and distance.[15] Kracauer's final use of the term *geometry* was as a metaphor to describe the rationalized spatial forms produced by modern capitalist society that embodied in material objects the troubling spirit of the age. The rationalization and scientific management of work, commonly referred to as Taylorism, expressed itself for Kracauer as the synchronized legs in dance revues and the animals in modern zoos, which moved "rhythmically" and formed "geometric patterns," like workers in the factory (K 5.2:403).[16] If these geometric patterns symbolized the intellectual torpor of contemporary German society in the 1920s, then they also offered an ideal point at which a specifically geometric critique of cultural products could make society aware of its defective rationality. For society today, which is all the more the product of mass and, now, digital culture, Kracauer's negative mathematics offers a theory of cultural critique that not only accommodates the perspectives of marginalized critics but also depends on their intervention to understand and change society.

The Axioms of Necessity: Geometry and the Impossibility of Pure Sociology

Kracauer's negative mathematics began with his first published book, *Sociology as Science* (*Soziologie als Wissenschaft*, 1922). Written within the decades after the founding of sociology as an academic discipline, the book investigates the possibility of a "pure sociology," an idea adapted from Husserl's attempt to model a transcendental, "pure phenomenology" on mathematics.[17] *Sociology as Science* asks and wrestles with the question: How is it "possible to understand social occurrences in their necessity" (K 1:9)? Can sociology come to the type of logically "necessary" conclusions that we find in mathematics, such as in Euclid's geometry? In essence, the book argues that the study of society could build on more than just "mere experience" and construct more than just "pseudo-laws," if it could only collect evidence such as "the foundational statements of geometry" by making "space, time, and the categories of the understanding" into the conditions of sociological knowledge (33, 36, and 44). As one expects and as the text admits, the attempt to found a pure sociology ultimately fails, because, as would become a mantra of the Frankfurt School, social life exceeds pure logical codification. And yet, even as a failed experiment, *Sociology as Science* makes a set of intellectual moves central to Kracauer's negative mathematics. On the example of geometry, the text establishes that there are modes of thought that can synthesize logic and materiality, even if sociology itself cannot. It also reveals the emergence of a metaphorics of space in Kracauer's thinking about mathematics that renders legible this potential synthesis of thought and experience. Finally, the book suggests a performative aspect in Kracauer's thinking and writing, which takes the intellectual enactment of the failed merger of mathematics and sociology as its interpretive success.

In a modern world in which, at least for Kracauer, life lacked meaning, *Sociology as Science* turns to geometry as an example of how thought can make logical judgments about the world of experience. In its first section, the book goes to great lengths to categorize the present day in contrast to a now-lost "epoch filled with meaning" (*sinnerfüllte Epoche*), a term borrowed from Lukács's *Theory of the Novel* (*Theorie des Romans*, 1921). Where meaning (*Sinn*) was immanent to life in a past "epoch," secularization and the rise of the modern natural sciences had cast the modern individual into "the cold infinity of empty time and empty space" (K 1:12). For Kracauer, a solution to the modern disjunction of life and meaning lay in the idea that there could be a set of logical rules that underpin and determine objects and events. "Necessity vanquishes chaos," he writes, and "the more necessity reveals itself to the knower, the more multiplicity congeals into a unity filled with necessity" (34). In the face of this negativity, the rift separating the materiality of experience and the "necessity" of logic, geometry offered a bridge: "A look at the geometric axioms, for instance, teaches that there are plenty of material relationships the experience of which is linked to compulsory thought. These axioms manifest themselves readily to any observation directed at them, they are intuitions, which cannot be proved further and cannot be derived from other knowledge, but rather themselves represent the original source of any experience building on them" (35). Take Euclid's geometry, which begins by positing self-evident "axioms," such as the notion that the whole is greater than the part. We "deem" this idea, which makes a statement about "material relationships" in the world, "worthy" of acceptance (as in the Greek *axīōma,*) because it is logically self-evident, even if we cannot prove it.[18] In *Sociology as Science*, the geometric axioms exemplify that, despite the modern separation of world and meaning, the link between our "material" experiences (e.g., chairs are greater than any of their legs) and "compulsory thought" ("the whole" is necessarily greater than "the part") had not been completely severed. Indeed, the previous quote curiously calls on the geometric axioms to blur the line separating experience and thought, materialism and idealism: we both know from "observation" and "intuition" that wholes are greater than parts and, at the same time, the knowledge that wholes are greater than parts conditions our experience as its "original source." The task that Kracauer sets for himself in *Sociology as Science* thus resides in determining if sociology, too, can produce knowledge about society along the lines of the synthesis of "experience" and "compulsory thought" in geometry. Even for the modern subject, mathematics held out hope that thought could find necessity, an immanent sense of meaning, in the apparent randomness of social life.

As the passage suggests, what is intellectually significant about geometry for Kracauer is not only how its axioms mix logic and materiality but also the fact that these axioms provide a starting point from which "compulsory thought" logically and irrefutably unfolds. Here *Sociology as Science* draws on the understanding of axioms and geometry offered by two of Kracauer's main intellectual interlocutors, Georg Simmel and Edmund Husserl.[19] For Simmel, the axioms of geometry may seem to be a logical necessity, but they, ultimately, depend on the human way of thinking, "our mode of perception." As he explains in *The Philosophy of Money* (*Die Philosophie des Geldes*, 1900), Euclid's geometry "has validity only in relation to specific physio-psychological organizations, their conditions of life and the furthering of their activity."[20] For example, the idea that the whole is greater than the part would have been for Simmel not an absolute logical property, but rather the accumulated result of how the human eyes perceived space throughout time. In contrast, geometry belonged for Husserl to the "sciences of the essence" along with pure logic, which do not depend on experience. In *Ideas: General Introduction to Pure Phenomenology* (*Ideen zu einer reinen Phänomenologie*, 1913), geometry exemplified the pure "thought-constructions" of the "sciences of the essence." The procedure of geometry "is *exclusively eidetic*" meaning that "from the beginning and in all that follows further it makes known no factual meaning that is not eidetically valid, in the sense that it could either be brought without mediation to primordial givenness (as being immediately grounded in essences of which we have primordial insight), or could be 'inferred' through pure consequential reasoning from 'axiomatic' factual meanings of this type."[21] The passage lays out what David Hilbert later called "axiomatic thinking" in mathematics: all knowledge in geometry is either "immediately" evident (and, hence, is an axiom) or it follows logically from an axiom.[22] For Husserl, the idea that the whole is always greater than the part was a logical necessity—along with any knowledge derived from this axiom—and would be true even without its confirmation through human experience. *Sociology as Science* paves a third way between Simmel and Husserl. For Kracauer, the geometric axioms depended on the materiality of experience, in that they were not the product of mental visualization, but their universality also conferred necessity and validity to any statement derived from them. In chaotic times, this feature of mathematics must have been attractive to Kracauer, in as much as it meant that a theory of culture could possibly derive logically valid theoretical positions from the meaning it found in material phenomena.

As a potential basis for pure phenomenology and sociology, mathematics provides *Sociology as Science* with the epistemological links between thought and experience, which the text codes in a metaphorics of space. As we saw with Scholem and Rosenzweig, mathematics illuminates for Kracauer the possibility of knowledge in the face of skepticism and relativism. As Kracauer explains: "The essential description of the simplest mathematical constructs has axiomatic significance for the entire area [*Gebiet*] of mathematics; it is the product of immediately evident observation [*Schauung*], which, for the named reasons, are linked with objectivity and necessary thought [*Denkzwang*]. Here, the relationship between mathematical statements and the statements of pure phenomenology becomes visible" (K 1:46–47). In this passage, we see the effects of Simmel and Husserl, as mathematics' "essential descriptions" mix logic and materiality. As with Simmel, the "simplest mathematical constructs" result not from abstract "intuitions" (*Anschauung*), but rather concrete "observations" (*Schauung*). As with Husserl, primordial constructs have "axiomatic significance" throughout mathematics, meaning statements that follow logically from "necessary thoughts" are also "objective" and "necessary." With this passage, however, I want to call attention to the spatial metaphors that *Sociology as Science* takes from Husserl and intensifies into a metaphorics: here, mathematics is an "area" and, throughout the text, mathematics and sociology each constitute a "field," a "manifold" (*Mannigfaltigkeit*), and a "continuum." The pure "essences" discerned by phenomenology form, as Kracauer explains, a "hierarchy, they smoothen, so to speak, into a truncated cone."[23] In *Sociology as Science*, the importance of these spatial metaphors is their propensity to reveal: in the previous quote, the "entire area of mathematics" renders legible the possibility of an intellectual domain, here represented by the geometric axioms, in which the messy world of experience and "necessary thought" interact. In a world in which knowledge and life seemed to lack a "secure, absolute foundation," these spatial metaphors held out hope for an "area" that accounted for both the materiality of experience and the logical structure of thought (34). In my analysis of Kracauer's version of cultural critique, I return to these spatial metaphors and their "intermediary area," as he later called it, in which material objects carry transcendental meaning and through which thought could intervene in the material world.[24]

In *Sociology as Science*, the metaphorics of space also provide the stage on which the argumentative failure of the text plays out. The characteristics of the space charted out by mathematics, geometry in particular, are its totality and regularity: the "mathematical manifold," Husserl writes, "determines completely and unambiguously on lines of pure logical necessity the totality of all possible formations in the domain."[25] The logical uniformity of "the field of mathematics" makes it, in the terms of *Sociology as Science*, "homogenous" (K 1:46). With the exception of the axioms, which mix experience and logic, all statements in a field like geometry—theorems about lines, propositions about triangles, and so forth—have the same logical form in that they all follow necessarily from (and only from) the axioms. For Kracauer, this homogeneity represents the epistemological advantage of mathematical reasoning: "the further one removes oneself from the axioms, the less transparent the figures become, whose necessary construction the pure ego seeks to grasp," he explains, "and it frequently takes many intermediary inferences and makeshifts, to cover the distance from the figures back to the immediately evident axioms. But despite its possible length, the path is always traversable" (48). Consider, for example, proposition 30 in Euclid's book 1: the proof that "straight lines parallel to the same straight line are also parallel to one another" draws only on the "intermediary inferences" of the previous-proved proposition 29 and a set of commonly accepted notions ("axioms").[26] For more complex systems of reasoning such as ethics, the spatial metaphor of distance renders legible why philosophers like Spinoza have often turned to mathematics as a ordered paradigm of thinking: it guaranteed that no matter how many intermediate steps it takes to reach a new proposition, the same "compulsory thought" that applied to the axioms will also apply to the conclusions drawn from them. However, the spatial "homogeneity" that makes up mathematics' epistemic advantage also pointed Kracauer to the epistemological difference between mathematics and sociology.

In contrast to mathematics, *Sociology as Science* argues that the space of phenomenology and sociology is "not homogeneous," illustrating the failure of the text's proposed construction of a "pure sociology" (K 1:49). Here, the text again follows Husserl, who rejected the endeavor to found a "geometry of experience" (*Geometrie des Erlebnisses*, 48).[27] In Kracauer's words, "the intention to collect experiences, whose necessity and generality are evident, in areas other than in mathematics" predictably and intentionally falls short: only "pure phenomenology in the strictest sense" ("the smallest part" of the "space tapering upward" of phenomenology) can possibly deliver "synthetic judgments a priori, which equal those of mathematics" (36 and 50). This failure lies in the fact that "the acts of consciousness" are not self-referential like the formal statements of logic and the ideal, spatial concepts of geometry, but rather depend on "the various modes of human community," they reference "things, values, etc." (51). For instance, acts of consciousness reference objects and ideas that exist beyond the world of consciousness, such as this book as a material object or the concept of a "book" as a product of certain society at a certain time. Kracauer's postulate of pure sociology thus remains an impossibility, as Inka Mülder-Bach explains, because of a "fundamental misunderstanding," in which "Kracauer holds sociology to a validity claim different than that of the sciences that study experience. He demands from it a type of 'objectivity' and 'truth' different than those that empirical research can provide and seeks to provide."[28] This assessment is correct, but this failure is the performative aspect of *Sociology as Science*, which also opens, exposes, and explores the seemingly self-evident incongruence between the world of experience and the world charted out by mathematics.

The spatial metaphors of fields, spheres, cones, and areas enact *Sociology as Science*'s performative failure, which suggests how negative mathematics functions as a metaphorics in Kracauer's thought. Throughout the text, these spatial metaphors put on display the mismatch of the homogenous mathematical "manifold" and the heterogeneous sociological "field"; they "demonstrate" (*erweisen*), as the text announces already in the introduction, how "formal philosophy" cannot encompass the "sphere of reality" that sociology seeks to describe (K 1:11). The assignment of these spatial metaphors is thus to expose, to reveal, and to render legible the impossibility of "pure sociology": they do the demonstrative work of showing that thought can neither exhaust the "field" of phenomenology (from material phenomena to the most general categorical essences) nor account for the infinite possibility of social occurrences. In contrast to the finite axioms of mathematics, sociologists would always have to add new sociological axioms to account for new social phenomena. For Kracauer, then, the metaphorics drawn from mathematics function differently than they did for Scholem and Rosenzweig. For Rosenzweig, the metaphorics of subjectivity and motion provided by infinitesimal calculus lent a language to limit points of the natural and spiritual world that philosophical and theological language could not describe. The metaphorics of space did not allow Kracauer to name the nameless, nor did he desire to; instead, they demonstrated and illuminated aspects of knowledge and experience, such as the mismatch of mathematics and sociology and the relationship between logic and materiality, that thought may have otherwise taken for granted and passed over as obvious and ubiquitous. The incongruence of mathematics and sociology may come as no surprise for readers today. But the way the text works out and puts on display the contradictions between thought and experience—here, formalized knowledge like mathematics and the social world—would become a signature move of the critical theorists.[29] Indeed, this performative dimension of negative mathematics informed Kracauer's cultural critique, as an analytic and literary technique to render legible the tensions between material and metaphysics as a mode of cultural intervention.

The sense of incommensurability that *Sociology as Science* produces in its comparison of mathematics and sociology characterizes much of Kracauer's writings from the interwar period onward. As Adorno writes, "incommensurability" was Kracauer's "central theme—which, precisely for this reason, hardly ever becomes thematic in his work."[30] Indeed, the sense of disjunction between materiality and logic ran through the works of first-generation critical theorists, such as Lukács, Horkheimer, and Adorno. In Lukács's *Theory of the Novel*, for instance, the novel takes over at the point where the alienation of modern life no longer fits the inherited literary form of the epic that presupposed the unity of humankind and nature. In Adorno and Horkheimer's later criticism of logical positivism, mathematical logic cannot provide, in Adorno's terms, the "unified interpretation of reality that it demands: namely, because reality contradicts it and because it itself is inconsistent."[31] A formal, mathematical system of knowledge can never account in full for the contingency and depth of lived reality. For Horkheimer and Adorno, this incommensurability caused them to dismiss mathematics and condemn any attempt to fit experience, the study of society, and cultural analysis into what was, in their eyes, an ill-fitting mathematical container. Kracauer, too, abandoned the idea of a mathematically "pure sociology." But *Sociology as Science* showed him that mathematics, in particular geometry, still held out hope for a method of analysis that could blend logic and materiality, allowing logic to intervene in the material world and experience to shape intellectual concerns. And, after *Sociology as Science*, it was space and spatial methods that continued to signify the possibility of combining thought and experience in Kracauer's writing. What remained was to find objects of analysis that occupied this liminal zone between the logic of critique and the materiality of modern life.

Projektionslehre: *Descriptive Geometry and the Detective Novel*

In the early 1920s, the idea that geometry dealt with negativity by bridging logic and materiality emerged in Kracauer's thought as a distinct method of cultural analysis and critique. As deployed in *The Detective Novel: An Interpretation (Der Detektivroman: Eine Deutung,* 1922–1925), a geometric method, "a sociological theory of projection [*Projektionslehre*]" allowed the cultural critic to read the material products of society, such as detective novels, as indicative of the metaphysical underpinnings of contemporary society and culture.[32] This turn to the analysis of mass culture and its products influenced early members of the Frankfurt School, but it also offered a methodological challenge to the seeming disjunction of meaning and life characteristic of modernity and exacerbated by the relativism of contemporary cultural analyses.[33] Projection countered, for instance, Simmel's associative and wandering intellectual style, which entailed for Kracauer a relativistic groundlessness that embodied "the fate of civilized humanity" in the age of capitalism (K 9.2:246). Where, in the terms of Kracauer's 1920 letter to Susman quoted at the outset of this chapter, Simmel "only describes" the cultural phenomena he analyzes, this "sociological theory of projection" allowed Kracauer to "explain," to "give [the] principles" that currently shape society.[34]

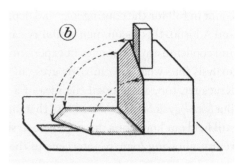

FIGURE 4.1. Projection, as depicted in Erich Salkowski's *Foundations of Descriptive Geometry* (1928), maps the three-dimensional side of the house onto the two-dimensional plane.

Beyond developing the metaphorics of space, the geometric method of projection served as the generative dimension of negative mathematics in Kracauer's thought. Run in reverse, projection provided a way to read the composition of mass-produced, aesthetic objects as indicative of the metaphysical principles that govern society, such as secular rationality. It also contained a crucial insight for cultural theory: the notion that such aesthetic objects can reveal these principles in as much as they reflected the logic of their creation in their material form.

Projection is a widely used critical term—from psychoanalysis to cinema studies—but it also carries with it a more specific, technical definition.[35] In descriptive geometry, a branch of mathematics developed in the eighteenth century by the French mathematician Gaspard Monge, projection refers to the mathematical technique of mapping one structure onto another.[36] The goal of descriptive geometry is not to expand on the *more geometrico* as a model of sound logical reasoning, but rather to provide practical and heuristic geometric tools to engineers, architects, and technicians "to imagine a convoluted figure intuitively."[37] The most common of these procedures is the spatial projection of a three-dimensional object onto the two-dimensional plane, such as the shadow cast by a building on the ground (fig. 4.1).[38] For architectural and engineering purposes, it is easier to manipulate mathematically a two-dimensional representation of a house—for example, to measure the height of the house or the angle of the roof—even though the process of projection sacrifices a dimension (the length of the house). Despite his apparent distaste for the subject, Kracauer's architectural studies would have required him to employ projection to draft technical drawings of the buildings and memorials he recounts producing in his autobiographical novel, *Ginster* (1928).[39] For Kracauer the cultural critic, projection offered a technique to represent the aesthetic products of mass culture in an intellectual space in which the societal and cultural principles governing their production could become readable.

As in *Sociology as Science*, the society that *The Detective Novel* analyzes is the modern, rationalized and secularized world, but *The Detective Novel* displaces its object of analysis from society itself to society's products, which it illuminates through projection. Best known for its chapter "The Hotel Lobby," *The Detective Novel* is a strange yet revealing text, neither a literary history nor a social history of the rise of the European detective novel. Instead, *The Detective Novel* offers, as Kracauer writes to Löwenthal, a "metaphysics of the detective novel."[40] It proposes and enacts "an interpretation" of "the idea, to which detective novels testify and out of which they are created [*von der sie zeugen und aus der heraus sie gezeugt sind*]: the idea of a thoroughly rationalized, civilized society" (K 1:107). What makes this "art of interpretation" possible is the aesthetic constitution of detective novels, the Enlightenment genre par excellence, in which plot, characters, and mise en scène formally reflect and contribute to the immanent triumph of rationality (the detective) over mystery (the crime). "Detective novels are not concerned with a representation of the reality called civilization that stays true to nature," Kracauer continues, "but rather, from the outset, with the exposure of the intellectual character of this reality; they hold a distorting mirror in front of that which is civilized, in which the civilized comes face to face with a caricature of its dreadful state of affairs [*Unwesen*]" (107). In the detective novel, modern society sees its image, however distorted, as unreflective, superficial, and uniform. Surprising about such a statement is how it locates the metaphysical keys to the "intellectual character" of an epoch not in world-historical events (such as the advent of first German Republic or massive economic inflation), but rather in a mundane yet mass-produced literary genre. Finding in the products of capitalist modernity a "mirror" of its "dreadful state of affairs" is the cultural-critical task undertaken by Kracauer's most famous feuilletons, such as "The Mass Ornament" and "Photography," and still practiced today as critical theory.

The geometric procedure of projection supplied the particular exegetical mechanism that made this mode of interpretation possible. The disjunction of life and meaning permeates the world of *The Detective Novel*, but projection offers a means of reestablishing a connection between this fallen epoch and a state of affairs in which logic and materiality coincide:

> When a person rejects the relationship [to the religious sphere], they de-realize themselves, but even still, apart from and outside the relationship, the features of the high sphere remain unshakably in effect. It is they that is meant with their displacement, which itself no longer means them, for in the cloudy medium things appear broken like the image of a stick dipped into water and all names are mangled beyond recognition. . . . Therefore, the ensnarled knowledge and behaviors of the lower regions have equivalents in the higher spheres; the message that such knowledge and behaviors bring depicts inessentially something essential. It is first their projection onto the very contents that they distort that makes the distorted images transparent: if their meaning is to be freed from the depths, then they are to be transformed until they reappear metamorphosed in the coordinate system of the high region of spheres, where they may be examined to determine their meaning. (K 1:109)

The passage intensifies the metaphorics of space, through the reference to "spheres," but also the spatial dimension invoked by a "projection onto" itself. In particular, projection allows the secularized individual to translate ideas, objects, and events in a rationalized society ("the lower regions") into their metaphysical plane of meaning ("the high sphere"). It deciphers the metaphysical "message" of "the ensnarled knowledge and behaviors of the lower regions." Here, the passage draws on an idea that Simmel proposed in the essay "On the Spatial Projection of Social Forms" ("Über räumliche Projektionen sozialer Formen," 1903), according to which physical space (e.g., a change in location of a capital city) often directly reflects sociological factors (a change in leadership).[41] But terms such as "theory of projection" (*Projektionslehre*, the technical term for studying geometric projections), "coordinate system" (the geometric, Cartesian plane), and "transformation" (a basic geometric operation) reinforce the mathematical dimension of Kracauer's criticism.[42] Indeed, we have already seen the connection Kracauer makes between his "thinking" and "higher mathematics" cited at the outset of this chapter; in that same letter, he continues: "I could begin my theory of knowledge, which constantly preoccupies me, as follows: given are two spiritual systems, X and X'. Which transformations must be made, in order to get from X to X'?"[43] What emerges in *The Detective Novel* is thus a hybrid notion of projection, blending Simmel with the spatial-geometric terms introduced in *Sociology as Science*.

In essence, Kracauer's "theory of projection" entailed a three-step interpretative procedure that reads these "distorted images" for their metaphysical implications: correspondence, projection, and examination. The first of these philosophical moves fleshes out what Kracauer means when the text claims that detective novels hold a "distorting mirror" up to "that which is civilized": it proposes the structure of and correspondence between the "spheres" of modern existence and metaphysical meaning that this interpretive relationship seeks to uncover and interpret. Even if Kracauer does not speak of Judaism and Christianity as directly as Scholem and Rosenzweig, his texts still borrow from a Judeo-Christian metaphysical framework, in particular from Søren Kierkegaard's theory of existence as "stages."[44] Accordingly, humans exist in three existential "spheres": the aesthetic (where individuals live in the sensuous present), the ethical (where we attempt to unify finitude and infinitude into a cohesive self), and the religious (in which, via a "leap of faith," we reconcile our paradoxical relationship to eternity).[45] "In the high sphere, according to Kierkegaard the 'religious' sphere in which the names disclose themselves," Kracauer writes, "the self stands in a relation to the high secret, which the relationship brings fully into existence. Word and deed, being and image move here right up to the outermost limit, what is experienced is real, what is known is of final human validity" (K 1:109). Objects in the lower sphere of a secularized, rationalized society thus have equivalents in the high sphere of a religiously oriented community, in which knowledge still possesses the "final human validity" that we saw ascribed to mathematics. By inverting geometric projection, we can read the products of the lower sphere in terms of their metaphysical "principles," which, through further interpretation, reveal the final principles ("the high secret") that contemporary society and its products follow.

The second exegetical step is the process of projection itself, which maps, as Kracauer writes, "the ensnarled knowledge and behaviors of the lower regions" onto their corresponding "contents" in the "high" sphere. Earlier I

FIGURE 4.2. Parallel projection is a form of projection in which projection lines, depicted as dotted arrows, are parallel to one another (examples from Salkowski's *Foundations*).

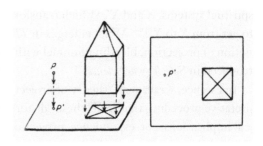

defined projections as the translation of three-dimensional objects (such as the cube topped with a pyramid in fig. 4.2) into their corresponding images on a two-dimensional plane (the bisected square) as dictated by a predefined perspective. The choice of projective perspective (cavalier or, here, parallel) determines the extent to which the representation is "distorted."[46] Kracauer's method of projection, then, is geometric projection in reverse; it reads the metaphysical shape of an object out of its corresponding, rationalized forms, flattened and contorted to the same extent that the Mercator projection distorts the earth's surface so that Greenland appears larger than Africa.[47] In Kracauer's text, the products of rational, civilized society—the detective novel, for example—are the two-dimensional figures. His version of projection then maps the detective novel into three-dimensional space, "the coordinate system" of the high region of spheres. In the terms of figure 4.2, the detective novel would be the bisected square, whereas its projected image, the cube topped with the pyramid, would be its "equivalent" in Kierkegaard's "religious sphere." Each element in the detective novel (the detective, the crime, and the novel's end) thus corresponds and can be interpreted as a metaphysical property (the human intellect, the divine secret, and the possibility of messianic reconciliation). The initial "distorted" image may be impoverished (without height in fig. 4.2).[48] It is thus the art of interpretation to transform this figure and reconstitute its corresponding shape in three-dimensional space in order to read an object's metaphysical meaning.[49] Reversing geometric projection turned the mathematical approach to negativity into a form of cultural analysis, bridging the divide between the detective novel as a mass-produced cultural object and its metaphysical sphere of meaning.

Take, for example, the iconic chapter on the hotel lobby from *The Detective Novel* as an instance of a cultural-critical method of projection. According to Anthony Vidler, the hotel lobby constitutes for Kracauer "the paradigmatic space of the modern detective novel," depicting the estrangement of the modern individual and epitomizing "the conditions of modern life in their anonymity and fragmentation."[50] As a mode of analysis, *The Detective Novel* projects the space of the hotel lobby into the religious sphere, reading it as "the counter-image [*Gegenbild*] of the house of God" (K 1:130). In the hotel lobby, the aimlessness and disassociation of the guests corresponds to and contrasts the "*assembly* and unification of the directed life of the community" in the church (131). As in the geometrical method of projection, Kracauer's analysis depends on the correspondence between the general characteristics of the hotel lobby and their equivalents in the church: the character and behavior of the guest and the sense of equality, the observance of silence, and air of mystery that dominates both venues.[51] For modern, secular society, comparing church and hotel lobby illuminates the social function of the latter. The fact that a community assembles in a church to be amidst a divine yet absent presence shows that the lack of this divine presence is the metaphysical principle of the hotel lobby, evinced by the estrangement and fragmentation of the hotel guests (133). In *The Detective Novel*, projection renders legible not only the metaphysical meaning of estrangement and fragmentation within the hotel lobby but also the lack of spiritual authority, or unified metaphysical meaning, in modern society as a whole.

The third and final philosophical step implied in Kracauer's methodological proposal is the "transformation" of the distorted figure until its "meaning" is legible in the "higher sphere." As the text continues in its main explanation of projection, previously cited:

> For these transformations it is important to keep in mind that the concepts and forms of life in the lower spheres have at least two meanings. On the one hand, what they mean corresponds to the conditions that govern the sphere that constitutes them. On the other hand, because the path of return is always traversable, and the decision remains open everywhere, they can house intentions, which are not proper to this lower sphere, but rather take on a really legitimate formulation only in one of the higher spheres. (K 1:109–110)

The metaphysical meaning of the projection of the "concepts and forms of life in the lower sphere" may not always be immediately obvious. Instead, the passage suggests, the critic must transform them interpretatively until they "take on a really legitimate formulation." Consider, for instance, how *The Detective Novel* interprets the sentimental conclusions of most detective novels. One may think such 'happy endings' meet the demands of popular literature in a society dominated by the material conditions of mass production and consumption. Interpreted as "the uncontested victory of *ratio*," however, the kitsch of detective novels' ending suggests, on a metaphysical level, the ultimate impossibility of a "messianic ending" (206). The almost obligatory resolution of the crime in the detective novel means that the Messiah will never come. Furthermore, the stipulation that "the path of return is always traversable" intonates that projection opens up for cultural critique a space between materiality—the hotel lobby, the experience and products of rational society—and the logically "homogenous" space of mathematics, where judgments derived from axioms carry the same analytic irrefutability of the axioms themselves. Through the projection and analysis of the logic evident in the material products of modern society, *The Detective Novel* distills and explores the metaphysical principles of modern society, which direct sociological analysis could not—as we saw in *Sociology as Science.*

Detective novels, as well as other products of mass culture such as films and dance revues, allowed for this type of analysis which otherwise evaded society as a whole, precisely because they were, as Kracauer often called them, "aesthetic creations" (110). Here *The Detective Novel* invokes the term "aesthetic" in special usage mentioned in the introduction to this chapter; studying detective novels is a matter not of "art works," but rather of perception as the physiological threshold between sensation and cognition. For Kracauer, the aesthetic products of modern mass culture adhered to lawful and logical "aesthetic principles of composition" that interweave "into a unity" revealing the "totality itself masked to those who bear civilized society" (118).[52] Detective novels reflect in their aesthetic-material composition the logic of their production, a "totality" that may otherwise remain hidden in that society. As a societal product, they "correspond" to and, as we have seen, render visible the "conditions" that produced them: producing, for example, the "unconditional victory of the *ratio*," while also being mass-produced by rational, "civilized society." Indeed, the often contrived and tortured prose of Kracauer's text, which often rearranges the word order of sentences in a way that makes readers' heads spin, serves as a stylistic reminder that a logic lurks behind both the detective novel and his analysis that requires mediation (i.e., projection) to render it legible. The geometric bridge between logic and materiality thus transforms in *The Detective Novel* into a geometric method of projection that enabled a "metaphysics of the detective novel," while Kracauer contemplated the "metaphysics of history" and proposed the need for "a yet unwritten metaphysics of film."[53] With the help of negative mathematics, what was impossible for the analysis of rationalized society became possible for the analysis of the rationalized creations of this society because their manifest rationality gave insight into the principles guiding their production.

The realization that the material products of a society reflect and reveal the logic of society was central to Kracauer's cultural critique and remains a core element of the critical project today. Reading the metaphysical consequences of aesthetic creations such as the detective novel—not to mention photography and film—became one of Kracauer's main intellectual objects, serving as a means, as he describes it in *History*, "to bring out the significance of areas whose claim to be acknowledged in their own right has not yet been recognized."[54] However, what goes unmentioned in this statement is the presumption that in order for the relationship between metaphysics and aesthetics to be critically binding, these creations—provided by literature, photography, architecture, and film—in some way replicate in their composition a set of discernable, perhaps even rational "principles," the search for which Kracauer found missing in Simmel. The nature of modernity thus becomes legible in its aesthetic products, because, as "products," they emerge out of the equally material and logical machinery of modern, mass production. For a critical apparatus such as the culture industry, which sees in the rationality of cultural products the deceptive and oppressive face of the capitalist rationalization of production, such an insight is indispensable. For Kracauer, whose critiques increasingly attended to the rationalized products of modern mass culture, the idea that the material products of society revealed the logic of that society held out the hope that critical analysis could not only diagnose but also, by diagnosing, intervene in the problematic form of reason governing contemporary life during the Weimar Republic.

The Geometry of Modernity: Rationality, Enlightenment, and the Mass Ornament

The method of geometric projection informed the analysis of Kracauer's best-known critical essay, "The Mass Ornament" and transformed in it into a political program for cultural critique. Published in 1927, the essay announces Kracauer's signature mode of reading modern culture, called "surface-level analysis" (*Oberflächenanalyse*).[55] "The place that an epoch occupies in the historical process," Kracauer proclaims, "can be determined more decisively from an analysis of its inconspicuous surface-level expressions than from the epoch's judgments about itself" (K 5.2:612). In the essay, the metaphysical space of projection shifts from theology to history—a shift that followed Kracauer's turn to a materialist-Marxist intellectual narrative in the middle of the 1920s.[56] Analyzing surfaces, however, is just another name for projection: the text reads rationalized spatial forms in modern life, which it calls "mass ornaments" on the example of the popular British dance revue, the Tiller Girls, as indicative of the contemporary capitalist "epoch" within the greater "historical process." Here we start to see the effects of negative mathematics in Kracauer's thought. By deciphering the meanings of the geometric patterns created by the Tiller Girls, projection rendered the stunted form of thinking at work in capitalist rationality legible, further laying bare the disjunction of materiality and meaning characteristic of modern life. At the same time, the blending of logic and materiality in this geometric analysis offered Kracauer a potential solution to this crisis through cultural critique as a means of interjecting into modern society the very form of reasonable thought that it so desperately lacked.

The relation of mathematics to capitalist rationality in "The Mass Ornament" reflects a larger trend in the early phases of critical theory, taken to its logical extreme in Horkheimer and Adorno's equation of instrumental reason and the catastrophes of the twentieth century. Let us consider an example from Georg Lukács's *History and Class Consciousness*, which was published a year after *Sociology as Science* and was highly influential for members of the Frankfurt School.[57] Lukács's *Theory of the Novel* greatly inspired a young Kracauer and Lukács's goal in his second work resonates deeply with Kracauer's critique of modernity. *History and Class Consciousness* explores rationalization in terms of reification, calling the latter "the central, structural problem of capitalist society." For Lukács, who adapts the concept from Marx, reification is the process by which "a relation between people takes on the character of a thing and thus acquires a 'phantom objectivity.'"[58] Capitalism depends on reification, because it allows human relationships to appear as abstract quantities, which we can calculate, equate, and exchange. According to Lukács, modern capitalism accelerates the process of reification by rationalizing not only "work-processes" through "mathematical analysis" but also "the economy" into "an abstract and, to the extent possible, mathematized system of formal 'laws' [*ein abstraktes, möglichst mathematisiertes Formsystem von 'Gesetzen'*]."[59] When I purchase a table from a store, for instance, interactions among humans—the labor that went into making the table, the process of shipping the table to the store, and so on—appear to me as an abstract quantity, the numerical price of the table. The problem with reification and rationalization lies in its transformative effect: "This rational objectification conceals above all the immediate—qualitative and material—character of things as things."[60] For Lukács, as well as for many first-generation critical theorists, mathematics served as the mechanism by which capitalist rationality covers up and neglects the qualitative features of human existence by rationalizing them into quantitative forms and relationships.

A similar line of reasoning relating mathematics to capitalist rationality is at play in Kracauer's diagnosis of modernity as the divergence of life and meaning. A decade before "The Mass Ornament," some of Kracauer's earliest known writings tend to see in mathematics the potentially detrimental, forced application of types of logical reasoning to objects and areas of study of qualitative character with which they are fundamentally incompatible—the implications of which can be seen in *Sociology as Science*. Regarding the types of thinking privileged by modern capitalism, Kracauer writes that "the technical gift of discovery, talent at organization, arithmetic dexterity, logical thinking, etc. become reified into products, whose value is not, for instance, inestimable, but rather allows itself to be directly expressed in numbers" (K 9.2:267). The reference to mathematics is oblique, but such obliqueness indicates its instrumental function: the concepts and operations of mathematics, such as "logical thinking," "arithmetic dexterity" and "numbers," only serve as tools to calculate, equate, and exchange "products" that are, in truth, "inestimable." The mathematical rationality of modern capitalism was thus an incomplete form of reason that privileged certain forms of thinking such as logical deduction and arithmetic reckoning but exhibited "a deep indifference towards the 'what' of things," human beings included (K 9.1:203). Mathematics mixed materiality and logic, but, when taken as ends in itself, transformed the complex and, ultimately, incalculable aspects of society into abstractions categorically unequal to and incongruous with the real people and things it represents. For Kracauer, this type of rationality characterized the problematic state of modern, mass capitalist society in early twentieth-century Germany, as represented by what he calls the mass ornament.

Whereas Horkheimer and Adorno's identification of mathematics with Enlightenment's relapse into barbarism caused them to dismiss the critical potential of mathematics, the bridge between materiality and logic in negative mathematics offers "The Mass Ornament" a way to expose, confront, and, potentially, overcome the pathological rationality of modernity. At its core, the essay introduces the concept of the mass ornament in order to analyze mass-produced cultural phenomena, such as zoos and dance revues, often associated with the United States and personified by the synchronized performances of the Tiller Girls.[61] Such phenomena are "mass," because they are mass-produced, have global appeal, and strive for mass popularity rather than elite artistic distinction; they are "ornaments," because of their decorative yet inessential function in society and the "pure assemblage of lines" that characterize them.[62] As in *Sociology as Science* and *The Detective Novel*, mathematics enters "The Mass Ornament" as a means of illuminating and understanding these modern creations. As the essay explains: "The ornament, detached from its bearers, is to be grasped rationally. . . . It consists of straight lines and circles, as are found in the textbooks of Euclidean geometry; it includes the elementary figures of physics, waves and spirals.

Discarded are the proliferations of organic forms and the emanations of spiritual life" (K 5.2:614). These sentences function on multiple interpretative levels. Referring to "Euclidean geometry," both sentences develop the metaphorics of space. With "geometry," the text does not mean Euclid's system of reasoning, but rather employs the geometric "straight lines and waves" as metaphors for the material expressions of capitalist rationality. The claim that we "grasp" (*erfassen*) such phenomena "rationally" suggests that only a mode of analysis that addresses the rationality of the mass ornament can adequately capture its implications and place on the metaphysical level of history, instead of dismissing it as a fad of lowbrow culture. Locating this place, we recall, is the goal of the essay and the second sentence hints at the historical stage indicated by the Tiller Girls: as in Lukács, it is a phase of capitalism that reifies "spiritual life," transforming "individual girls" into "irreducible complexes of girls, whose movements are mathematical demonstrations" (612). Finally, through the haptic dimension of the term *erfassen* (literally, "to grasp"), the first sentence implies that rationality, the rational analysis of the mass ornament, has the power to intervene in and, potentially, alter the course of history that produced it.

In essence, "The Mass Ornament" projects the mass ornament into a metaphysics of history in order to read the historical stage occupied by Germany during the Weimar Republic, a capitalist society near the apex of industrial expansion in a liberal democracy. It thus draws on not only the methodology of projection in *The Detective Novel* but also its presupposition linking rational society and its cultural products. "The Mass Ornament" begins this process by defining the guiding principle shared by the Tiller Girls and the society that produced it: "[Both are] designed according to rational principles, which the Taylor system simply takes to its ultimate conclusion. The hands in the factory correspond to the legs of the Tiller Girls. Beyond just manual talents, psycho-technical aptitude tests attempt to calculate even spiritual dispositions as well. The mass ornament is the aesthetic reflex of the rationality to which the ruling economic system strives" (K 5.2:615). Projection relates the mass ornament as an "aesthetic reflex" to the capitalist "rationality" of modern society that created it. As before, the passage uses the term "aesthetic" here to refer to both the material and cognitive dimensions of the mass ornament. The Tiller Girls, for example, occupy this aesthetic liminal zone, a phenomenon produced for mass entertainment that reflects, in the visual form of geometric dance patterns, the logic of mass production in the "factories." This logic is the "rationality" proper to capitalism ("the ruling economic system"), the scientific management of the "Taylor system." Such rationality also indicates, as discussed at the outset of this section, a troubling form of thought that replaces the deeper dimensions of "spiritual dispositions" with the mathematical calculations of "aptitude tests." The text reinforces this sense of stunted reason by calling the mass ornament a "reflex"—the Tiller Girls are not the conscious creation of the reflective intellect, but rather reflect mass production designed for mass consumption.

As the next step in the process of projection, "The Mass Ornament" defines a new metaphysics of history in which its analysis of the Tiller Girls allows readers to situate their own contemporary moment. In "The Mass Ornament," history is no longer a narrative of a divine "meaning" (*Sinn*) that has abandoned the world. Instead, the text adapts the thesis from Max Weber that modernity constitutes the horizon of a larger process of "demythologization." This metaphysics not only eschews a narrative of a fallen present vis-à-vis a past Golden Age but also anticipates Horkheimer and Adorno's definition of the Enlightenment project: "The process of history is a battle fought out between weak and distant reason [*Vernunft*] and the forces of nature that ruled over heaven and earth in the myths" (K 5.2:616).[63] History progresses to the extent that "reason" vanquishes "myth" as the basis of human life. For Kracauer, this concept of reason was a privileged form of Enlightenment reason (*Vernunft*), the "reason of fairy tales" (617). In the text, the term "fairy tales" represents not the irrationalism of folklore, but rather a vision of the world in which "truth" and the human, such as a Snow White, unilaterally win out over the mythic forces of evil, such as the queen. The rational, but not fully reasonable, mass ornament projected by the text thus indicates the place of contemporary society within this metaphysical process. "The capitalist epoch is," as "The Mass Ornament" claims, only "a stage on the path to demystification" (617). Capitalist rationality is reason, but one that the privileging of mass production and mass consumption over the concerns of humans themselves has arrested into partial, "murky" reason. We may object to the simplicity of such a linear theory of history, but

Kracauer's theory of historical progress is not what is significant here. What I want to draw attention to instead is how mathematics represents here not the blind tool of capitalism, but rather the method that renders legible the idea that capitalism only constitutes a phase of reason's development. If capitalism is only a historical "stage" (note again the spatial metaphor), then perhaps there is hope that thought, properly configured, could surmount it.

The notion that society could potentially push past the contemporary phase of "murky reason" was the moment where negative mathematics opened a critical window in Kracauer's thought. To recap, the critical point on which "The Mass Ornament" has been building up to here holds that the manifold problems of capitalist society stem, at least in part, from a shift in societal focus. Capitalist society privileges abstract quantities (such as production, exchange, and profit) over humans and the qualities of human life. In Kracauer's terms, natural and spiritual forms, humanistic phenomena such as community and personality, and, perhaps most significantly, humans as such disappear (as exemplified by the Tiller Girls' transformation into synchronized shapes, into a realm where "what is demanded is calculability" [K 5.2:614]). For Lukács, the solution to this problem lay in a return to the classicist aesthetic program of Friedrich Schiller and its notion of play, which potentially salvages life "from the deadening effects of the mechanism of reification."[64] "The Mass Ornament," however, rejects refuge into high art in the same way that Kracauer had earlier criticized the recourse to theology in Rosenzweig and Buber's Bible translation.[65] Such proposals were unfit to make this intervention, Kracauer argued, because they neglected the material reality of modern society in favor of idealist and outmoded religious solutions (623). Instead, the solution came in the form of confronting the problem that capitalism, as "The Mass Ornament" claims, "rationalizes not too much, but rather *too little*" (618). This enigmatic phrase was a call to fashion ways to reinstate reason and reflection back into a society governed by "murky reason." With a foot in this world of rationality, mathematics and projection allowed cultural critique to make this intervention, adding to the rationality of society not only by calling attention to it but also by critiquing it.

For Kracauer, the political imperative of not only geometric projection but also cultural critique as a whole lay in bring society face to face with its pernicious rationality and, through this rational confrontation, promoting the progress of reason. As Johannes von Moltke shows, photography and film were one way Kracauer envisioned this confrontation. As in texts such as "Cult of Distraction" ("Kult der Zerstreuung," 1926), modern cinemas fail to fulfill their political assignment when they cover up the distraction of mass media; instead, they must expose it in order to reveal to the masses their own distracted and disintegrated state.[66] Interpreting the "distorting mirror" in *The Detective Novel*, projection was another mode of cultural critique. "The Mass Ornament" thus ends on a self-reflective note: "[The 'process' of societal change] leads through the center of the mass ornament, not away from it. It can move forward only if thought encompasses nature and produces the human as he is constituted by reason. Then society will change. Then, too, the mass ornament will disappear and human life will assume the traits of that ornament into which it develops, through its confrontation with truth, in fairy tales" (K 5.2: 623). In practical terms, this passage imagines a "confrontation" of society with rationality as taking place through the enjoyment but also the analysis of the products of mass culture—the Tiller Girls, film, and photography, to name a few. Forcing society to reflect on its perversion away from "human life" meant an increase in "reason" and thus bore the potential of advancing reason and spurring society to reshape itself around humanity and not capitalist rationality. What is significant in this passage is not just its spatial metaphor ("through the center") but also its performative dimension—the same performative aspect of Kracauer's analysis we first saw associated with geometry in *Sociology as Science*. In a text such as "The Mass Ornament," the method of projection enables cultural critique, the analysis of the mass ornament, as a means of staging this confrontation in its readers. Bridging materiality and logic, the geometric approach to negatvity revealed an analytic technique through which the reason of cultural critique could intervene in and, potentially, remedy the rational material fabric of life in interwar Germany. Alongside the geometric method of projection, the metaphorics of space associated with it also offered literary techniques to try to confront readers with this rationality through the text itself.

In "The Mass Ornament," one sees not only the similarities but also the stark differences between Kracauer and Horkheimer and Adorno. In Kracauer's Weimar-era text, a form of cultural critique that draws on geometric projection still held out a utopian hope that the confrontation of reason with itself could advance the progress of history. In the intellectual climate of postwar Germany, it could only have seemed like illusionary optimism. For texts such as *Dialectic of Enlightenment* and *Eclipse of Reason*, not only mathematics but also the Enlightenment ideal of reason itself symbolized "the germ of the regression which is taking place everywhere today."[67] And yet Kracauer's vision of criticism exhibits a belief in critique's relationship to emancipation that persisted in Horkheimer and Adorno's notion of criticality and that persists in critical theory today. The geometric method of projection facilitated an analysis of society that was critical, not only because it allowed for a way of reading in the products of mass culture the social and economic principles that brought them into existence. This analysis was also critical, because it allowed the marginal figure of the cultural critic, someone with an oblique relationship to power and the production of culture, to intervene in and work toward the historical advancement of reason. Negative mathematics enabled Kracauer's Weimar-era feuilletons to take this political assignment of critique seriously, staging a confrontation of society with rationality through the products of mass culture that it analyzed and the critical forms that it employed to analyze them.

Natural Geometry and the Aesthetics of Theory

As geometric projection in "The Mass Ornament" turned into a political mode of cultural critique, projection and the metaphorics of space became aesthetically operative in Kracauer's interwar explorations of the rationalized spaces of the modern city. In these texts, the performative element we saw first in *Sociology as Science* and *The Detective Novel* took on the political charge of Kracauer's cultural critique as a literary strategy in texts such as "Lad and Bull" ("Knabe und Stier, 1926), "Two Planes ("Zwei Flächen," 1926), and "Analysis of a City Map ("Analyse eines Stadtplans," 1926). Indeed, the postwar publication of Kracauer's Weimar-era essays in *The Mass Ornament* (*Das Ornament der Masse*, 1963) made the connection of these three texts to geometry explicit, as they appeared together under the title "Lead-In: Natural Geometry." "I could imagine," Kracauer explains to the editors at Suhrkamp in his plan for the volume, "that these three little pieces could form, in terms of mood, a good prelude. Because of their geometric character, they resonate well with the word 'ornament' in the main title."[68] In this section, I explore the point of transference where geometric projection and the metaphorics of space became a "natural geometry" in Kracauer's cultural critique, which takes on a "geometric character" by rearranging the language and textual space of critique itself as a projection of rationalization. We can call this Kracauer's aesthetics of theory, in as much as these texts enact cultural critique, staging on the level of literary form the projection of rationality into the metaphysics of history. The texts of "Lead In: Natural Geometry" present this aesthetics of theory as an active, compositional strategy, which consciously rationalized aesthetic, textual space in order to promote the progressive confrontation of society with rationality called for by Kracauer's more programmatic essays published during the Weimar Republic, such as "The Mass Ornament."

As with geometry itself, the idea of "natural geometry" bridged materiality and cognition, suggesting a link between perception and our material experience of the things around us. Far from the rigid axiomatic systems of Euclid or Hilbert, the term natural geometry refers in mathematical and philosophical discourses to conceptual and pedagogical approaches to geometry that emphasize its visually intuitive dimension. The Italian mathematician Ernest Cesàro, for instance, proposed a geometry that avoids the use of nonessential coordinate systems in his book *Lectures on an Intrinsic Geometry* (*Lezioni di geometria intrinseca*, 1896) published in German as *Vorlesungen über die natürliche Geometrie* (1901).[69] The term, however, stems from the theory of perception proposed by Rene Descartes in his *Optics* (1637).[70] For Descartes, natural geometry (*géométrie naturelle*) designates our innate ability to comprehend and navigate the three-dimensional space that surrounds us, the possibility of which is conditioned by "the shape of the body of the eye"—that is, the physiognomic arrangement of the human sensory organs.[71] In Descartes's example, we consider a blind person who "sees" with the aid of the sticks, *AE* and *CE* (fig. 4.3).[72] Given "the relation of the eyes to one another," Descartes explains, "our blind man, holding the two sticks *AE*,

FIGURE 4.3. In the Illustration of the "blind man" in Descartes's *Optics* (1637), sensation in the hands (points *A* and *C*) corresponds with objects at the ends of the sticks (*A* with *D*, *C* with *B*). Image courtesy of University of Michigan Library, Special Collections Research Center.

CE, of whose length I am assuming that he is ignorant, and knowing only the interval which is between his two hands *A* and *C*, and the size of the angles *ACE*, *CAE*, can from that, as if by a natural geometry, know the location of the point *E*."[73] This "natural geometry" mediates between empirical experience in the form the haptic resistance provided by "the medium of the stick" and cognition in the form of knowledge of "the location of the point *E*" in real and abstract space.[74] If the geometric axioms, for Kracauer's pure sociology, once mixed material judgments with necessary thought, then natural geometry mixed the physical experience of modernity with cultural critique, that is, the materiality of modernity with its reflection in thought.

While we have seen this combination of materiality and cognition in the term *aesthetics*, the idea of natural geometry takes the entwinement of experience and logic a step further. Scholarly work on Kracauer has long recognized the materialist dimension in Kracauer's thinking, especially his film theory. As Miriam Hansen puts it, Kracauer espoused a "program of cinematic materialism" that draws as much on a neo-Marxist notion of materialism as a belief in the tactile effects of film on its audience.[75] According to Descartes, such tactility is a property of visual perception as such: "As to position, that is to say the direction in which each part of the object lies with respect to our body, we perceive this with our eyes in the same way as we would with our hands." Where previous theories of perception focused on the impression left on the mind by objects, "this knowledge does not depend," Descartes claims, "on any action which proceeds from the object," instead the active mind comprehends the object through experiencing it.[76] The idea of natural geometry presaged Kant's Copernican revolution, holding that the cognizing mind plays an active and sovereign role in forming and discovering the natural world.[77] Natural geometry also suggested that the material dimension of cultural critique could have a cognitive effect on its reader, beyond intellectual comprehension. The materiality of critique— the composition of the text, the medium in which readers encountered it— thus bore the potential to work like film, as Kracauer later called it, in the intermediary zone between experience and thought. This is not to say that texts like those contained in "Lead-In: Natural Geometry" are cinematic. Rather, they share their political imperative with film by reproducing and, thereby, confronting their readers with capitalist rationalization on the level of literary form and through a mass-produced textual medium, the newspaper.

The three texts that constitute "Lead-In: Natural Geometry" transform the metaphorics of space into a critical, literary form. Instead of discussing the position of contemporary society in the metaphysics of history explicitly, these texts project this place for Kracauer's audience through the texts' rationalized style. Where a text such as "The Mass Ornament" analyzes and projects a material phenomenon (the Tiller Girls) as the current state of reason (the rationality of capitalism), a text such as "Analysis of a City Map" renders this rationality legible through a detached, cool yet also impressionistic, bird's-eye examination of the *faubourgs* and city center of Paris. The effects of rationalization become particularly evident in Kracauer's depictions of the otherworldly urban spaces of Marseille's old harbor and so-called "Place de l'Observance" in "Two Planes."[78] As invoked by the mathematical term "plane" (*Fläche*), this text is a literary projection that maps the rationalized, two-dimensional "plane" of "The Bay" ("Die Bai") and "The Quadrangle" ("Das Karree") into an aesthetic, three-dimensional space. "The Bay" leads readers through the Old Port of Marseilles, but its aesthetic core comes in the opening sentences:

> Marseilles, a dazzling amphitheater, arises around the rectangle of the old harbor. Each one like the next, rows of facades fringe the three shores of the area paved with sea, whose depth cuts into the city. Across from the entrance to the bay, the Cannebière, the street of all streets, breaks into its smooth luminescence, carrying the harbor further into the city's interior. It is not the only connection between the soaring terraces and this monster of a square, from whose foundation the neighborhoods rise like the jets of a fountain. To the square point churches as the vanishing point of all perspectives, and the yet uncovered hills face it as well. Such an audience has rarely ever been assembled around an arena. If ocean liners were to fill the basin, their trails of smoke would drift to the most remote houses; if fireworks were to be set off over the plane, the city would be witness to the illumination. (K 5.2:468)

The "rectangle" that begins "The Bay," like the "square without mercy" that ends "The Quadrangle," immediately underscores the "geometric character" of these two texts. As Andreas Huyssen notes, the "naturalness" of the "geometry" in a passage like this is ironic, in that abstract rationality permeates the city and viewer in the same way it permeates the "legs of the Tiller Girls" in "The Mass Ornament."[79] Indeed, the passage stages, as I explore in the remainder of this section, this rationality by objectifying verbs, eliding humans, and, ultimately, rationalizing the grammatical space of the sentence itself. Through an aestheticized analysis, Kracauer's method of projection enabled texts such as "The Bay" and "The Quadrangle" to force readers to confront and reflect on the pervasive rationality of modernity.

The projection of rationality and the confrontation it hopes to effect in its readers occurs most noticeably in the interplay among space, shapes, and the observer. "Marseille," the first sentence tells us, "arises around the rectangle of the Old Port [*baut sich um das Rechteck des Alten Hafens auf*]." Like Descartes's active viewers whose cognitive activity codetermines their surroundings, the act of reading literally builds up "Marseille" out of and around the "rectangle of the Old Port" as if describing a photograph. Here the text combines a reflexive verb (*sich aufbauen*) with an inanimate object, affording an action ("to build"), strictly speaking, incongruous with the designated subject ("Marseille"). This formulation is accordingly difficult to render with precision in English: Marseille really does not "arise" around the square of the old harbor and it has not "been built around" the Old Port, but rather: "Marseille . . . builds itself up around the square of the Old Port." The experience of Marseille is not the product of humans, but rather builds on itself by virtue of the shift in the viewer's gaze. A relatively short text, "The Bay" repeats this construction: "the splendor" of the "sail-fishing industry" "has lost its luster [*hat sich abgenutzt*]"; "the streets dead-end [*laufen sich tot*] on its banks" (K 5.2:469). Such constructions anthropomorphize aspects of the city as the active reader of "The Bay" observes and thus animates its construction in tandem with the act of reading. At the same time, the text displaces the agency of the narrator to that of the readers. The style of depiction results not from the author's active retelling of his or her physical discovery of the harbor and promenades of the city, akin to Edgar Allen Poe's nameless protagonist in "The Man in the Crowd" (1840).[80] Instead, for Kracauer, narration occurs in the consciousness of the reader, in which the city generates itself around the old harbor, as if projecting a city map or an aerial photograph into a higher "coordinate system." On the level of form, "The Bay" confronts the rationality of the reader with the rationality of the city, creating an aesthetic link between reader and city but also underscoring the seeming autonomy of rationality in modern urban space.

The fact that its population is almost totally absent in Kracauer's depiction of Marseilles further underscores this sense of the autonomy of rationality. Near its end, the paragraph previously quoted reintroduces a human element to the otherwise empty streets: "Such an audience has rarely ever been assembled around an arena." While the term "audience" carries the connotation of human subjects—although, significantly, not individuals—the context of the preceding sentence informs readers that "such an audience" here refers to "the churches" and "the yet uncovered hills," and even the observer, who populate the "amphitheater" surrounding and observing the bay. Whereas "The Quadrangle" relies on people such as "the dreamer" and an "observer," "The Bay" erases the distinctions between individuals as in "The Mass Ornament," referring to "the human fauna" or "the masses of peoples, in which the people of different nations disappear." The text even erases individual authorship in its published form, as the "The Bay" (along with "Lad and Bull") first appeared in the *Frankfurter Zeitung* under Kracauer's generic staff byline, "raca."[81] The noticeable lack of humans in the text reflects the feature of Kracauer's film theory that von Moltke calls his "curious humanism": the emancipatory promise of cinema lies less in its depiction of humans on the screen, but in the interaction between objects on the screen and humans in the audience.[82] "The Bay" attempted to achieve this relationship between reader and text. In particular, it allows its reader to mingle with the rationalized object of the city, with individuals reduced to "fauna" and "masses" in the same way that the mass ornament rationalized "individual girls" into "complexes of girls." Eliminating qualities like "community" and "nationality," the text projects for the reader the substitution of rational objects for humanity, suggesting that the basis of Marseille is not a reason centered on humans, but instead capitalist rationality.

As "The Mass Ornament" hoped to accomplish a projection of rationality through its critical examination of the Tiller Girls, "The Bay" further stages this confrontation by rationalizing the space of the sentence. As "The Quadrangle" calls attention to rationality's domination of the subject, as Huyssen explains, the second sentence of the previous passage accentuates the traces of rationality in "The Bay." "Rows of facades, each like the next, fringe, on the three shorelines, the sea-covered square, whose depth cuts into the city [*Den meergepflasterten Platz, der mit seiner Tiefe in die Stadt einschneidet, säumen auf den drei Uferseiten Fassadenbänder gleichförmig ein*]." Here the sentence separates its subject ("*Fassadenbänder*") and direct object ("*Den meergepflasterten Platz*") from the verb ("*einsäumen*"). The emphasis of "rows of facades" and "the square paved with sea" created by this distance signals the text's own rationalized and calculated style. Not grammatically incorrect in German, this technique appears repeatedly not only in "The Bay" ("in the puddles the sky is pristine [*rein steht in den Lachen der Himmel*]" for example) but also in texts such as *The Detective Novel*.[83] Furthermore, the relative clause ("whose depth cuts into the city") and the spatial adverbial phrase ("on the three shorelines") reinforces the sense of interaction between material form and intellectual content, "cutting" into the flow of the sentence, enacting its content through its syntactic form. Within the calculated space of the sentence, the text forces readers to recognize not only the rationalization of space in Marseille but also of the textual space itself. One of the effects of emphasizing the spatial arrangement of sentence structure is to draw the readers' awareness to rationality and stimulate critical reflection about the rationalization of urban space and the everyday products of modern society, such as the newspaper in which "The Bay" appeared. This intermingling of materiality and logic constitutes the final consequence of the metaphorics of space that we have seen emerge throughout this chapter: the bridge between materiality and logic in geometry became an aesthetic strategy for cultural theory, mixing the materiality of literary style with the logic of cultural critique. Geometry showed cultural theory its aesthetic side, as texts such as "The Bay" literally lead readers through rationalization. By accentuating rationality formally, negative mathematics allowed Kracauer to compel his readers to think through rationality in the hope that beyond it lay the reason promised by the Enlightenment.

In printed form, "The Bay," the other texts that make up "Lead-In: Natural Geometry," and Kracauer's Weimar-era essays more generally reflected the projection of rationality in not only their aesthetic but also their material form. Most appeared in the feuilleton section of the newspaper, the *Frankfurter Zeitung.* In contrast to Scholem's private translations of a holy text, the Book of Lamentations, Kracauer's newspaper texts embody the logic of mass production and consumption that they often worked to expose, as quickly written, distinctly modern products. For instance, "The Mass Ornament" was only one of almost two hundred essays, film and book reviews, and reports that Kracauer published in 1927, equivalent, on average, to more than one publication appearing every other day.[84] Moreover, they appeared in a medium, a newspaper, which was synonymous with the rise of modernity and the rationalization of knowledge, indicative, for Benjamin, of the withering of the type of knowledge that he upheld as "experience" (*Erfahrung*).[85] As feuilletons, the texts in which Kracauer unfolded core elements of his theory of modernity are often "aesthetic" in the same sense as detective novels are: they reflect and discuss contemporaneous political and cultural developments, and maintain a pretension to literary styling, mixing materiality with thought. If, for Kracauer, the Tiller Girls created a mass ornament, then the form of the feuilleton itself, as understood for example by Karl Kraus, corresponded to a "literary ornament," carving out its own consumable literary genre for "the crowd" (*der Pöbel*).[86] Kracauer's texts, especially as they appeared during his tenure at the *Frankfurter Zeitung*, thus participated in the same process of rationalization whose more general cultural effects they simultaneously sought to lay bare. Participating in the same discourse his essays criticize is not the contradiction, but rather the critical imperative of negative mathematics in Kracauer's work, which sought to illuminate—indeed, to project—the contemporary state of society through such rationalized forms as an impetus to advance it.

For Kracauer, natural geometry held out hope that cultural critique could have a material, intellectual, and corrective effect on the society it critiqued. As geometric projection and the metaphorics of space turned into a literary style, Kracauer's text performed a political assignment, calling attention to the rationalization of space through the rearrangement of sentence structure, ascribing reflexive actions to objects, and the elimination of humans amidst the urban landscape. Kracauer's negative mathematics thus offers tools to cultural theory today, in particular the idea that the medium of critique can help render its message legible. This aspect of negative mathematics foreshadows and reemphasizes the interweaving of form and content in Horkheimer and Adorno's vision of critical theory, which underpins their prose.[87] For Kracauer, however, how geometry made cultural critique a space for cultural intervention was the emancipatory element of negative mathematics, which continues in the critical project today. If society necessitated the projection of its shortcomings onto a metaphysical level, then it also necessitated cultural critics, those who, based on their marginality, could observe mainstream society from the sidelines in order to change society. As cultural critique, negative mathematics thus included, as Kracauer called it in 1947, the "Jewish contributions in our era," which works "to dissolve all the elements that obstruct the breakthrough and fulfillment of reason." Through "centuries of migration, exile, and eternal adjustment," Jewish thinkers had also come to occupy the liminal zone between material permanence and transcendental reason, reality and thought, enabling them to "spread Enlightenment with a flashlight."[88] Negative mathematics enabled Kracauer as a cultural critic to spread Enlightenment by aestheticizing theory. For cultural theory today, negative mathematics suggests that the continued ability of cultural critique to enlighten depends on not only the objects it analyzes but also the aesthetic and material forms through which this analysis occurs.

Material Logic and the Politics of Cultural Critique

This chapter has traced the transformation of the metaphorics of space and the method of projection out of Kracauer's writing on geometry and into an aesthetics of theory—cultural critique that attempts to alter society by performing its critique through writing. Kracauer's essays, which this aesthetics of theory informed, and which appeared mainly during the Weimar Republic, thus point toward an alternative approach to the merger of the strict rationality of mathematics and cultural theory, unilaterally rejected by Horkheimer and Adorno. Kracauer's union of "higher mathematics" and "thought" by no means sought the totalizing mathematization of thought that in logical positivism, according to Horkheimer and Adorno, threatened to eliminate the critical and historical faculties of philosophy. In contrast, we see through Kracauer how negative mathematics mediates between materiality and logical necessity, enabling the critic to read the metaphysical dimensions of society in its material products and intervene in it through the materiality of critique. Kracauer's negative mathematics reflects the inclusivity of the theories of history, tradition, and knowledge that negative mathematics opened up for Scholem and Rosenzweig. Indeed, making the cultural critic an arbiter of social change, Kracauer's negative mathematics not only opens up the critical project to voices on the margins of society but also indicates that cultural critique depends on them. Moreover, by showing the potential political efficacy of the aesthetics of critique, Kracauer's negative mathematics recommends ways of putting this more capacious vision of critical theory into practice.

Informed by negative mathematics, the mix of logic and materiality in Kracauer's theory of cultural critique includes marginalized cultural perspectives and helps reincorporate them into the critical project. In contrast to Scholem's history of discontinuity or Rosenzweig's messianic theory of knowledge, this sense of inclusion in Kracauer's negative mathematics functions less on a theological than on a social level. In the same way that the liminal space of geometry blended materiality and logic, the vision of the cultural critic made possible by negative mathematics occupies a socially liminal space. Such a cultural critic would participate in the materiality of culture, while observing its logic from a critical distance attached to their social status as outsiders, strangers, and observers. In this vision of cultural critique, the ability to observe, analyze, and intervene in contemporary society would follow neither from the pronouncements of academic philosophers nor from the cultural critics that are part of the majority population of that society. Instead, this ability lies with cultural critics who have one foot inside and one foot outside society: the "homeless souls" of modernity, the exiles, the German Jews. Negative mathematics in Kracauer's thought turns the cultural critic into what Adorno later called "the dialectical critic of culture," who both "participate[s] in culture and [does] not participate."[89] In this regard, Kracauer's vision of cultural critique informed by negative mathematics circumvents Adorno's claim that cultural criticism depends on and, thus, only perpetuates the economic and cultural factors that it seeks to criticize. Kracauer's negative mathematics thus points us to a mode of cultural critique in the present as a form of self-understanding in contemporary society—drawing on a critical confrontation with its shortcomings—that depends on paying attention to the critical voices society may otherwise push to its margins. They, and perhaps only they, can catch glimpse of the conspicuous logic of society hidden within its material products.

Furthermore, negative mathematics bears the possibility that cultural theory and critique could alter society instead of merely describing, analyzing, and criticizing it. Drawing on the process of projection that Kracauer found in geometry, such a theory of culture would take the space of cultural critique as a mode of performing reason with the intent of effecting social change. For critical theory in the present, negative mathematics recommends techniques of making its cultural intervention through the aesthetic materiality of critique, to borrow a term from *Sociology as Science*, through a "material logic."[90] In particular, negative mathematics offers techniques of manipulating the form of an argument—through sentence construction, verb choice, and the presence of humans in the text—in order to draw attention to its critique (for Kracauer, of rationality) on an aesthetic level. These techniques suggest that the material manifestation of thought may provide one way that cultural critique could help put into practice a theory of history as discontinuity and a messianic theory of knowledge. In this regard, cultural critique could point to rupture and absence by manipulating not only the style of writing but also the material forms that communicate it, such as through print or, now, through digital technologies. Kracauer's negative mathematics thus puts renewed emphasis on one of critical theory's central tropes: the performativity of critique. Indeed, a "performative contradiction," as Habermas calls it, underlies a text such as *Dialectic of Enlightenment*, which works through the same sense of reason whose innermost contradictions its authors sought to expose.[91] Even if we may doubt the utopian hope of forwarding the progress of enlightenment, negative mathematics underscores the political urgency of the performative element of cultural critique as a means of projecting and calling attention to the contradictions of society. Through negative mathematics, cultural critique can find new ways not only to reflect on our society through philosophy but also to intervene and alter a society through the material and aesthetic dimensions created by the digital age.

Ultimately, Kracauer's later writings on film and history turned away from the generative potential that his Weimar-era essays found in negative mathematics. In his final work *History: The Last Things before the Last* (1969), mathematics provides the foil against which the task of the historian unfolds. "While the establishment of the world of science, this web of relationships between elements abstracted from, or imposed upon, nature, requires mathematical imagination, rather than, say, moral ingenuity," Kracauer explains, "the penetration of the historian's world which resists easy breakdowns into repeatable units calls for the efforts of a self as rich in facets as the human affairs reviewed."[92] For Kracauer, the study of history had to mediate between the contingency of its subject matter and the rule-bound logic of the natural sciences, sacrificing its study of the past to neither the one nor the other. Nonetheless, the type of cultural critique enabled by negative mathematics in Kracauer's Weimar-era essays must resonate with those of us who live in a world of new media, one ever more mediated and controlled by computers and other digital technologies. To be sure, we live in a world governed by a different type of rationality than the one discussed and performed in Kracauer's feuilletons. The rationality of this new form of capitalist society, however, is no less murky or any less in need of illumination than the rationality that accompanied the advent of mass culture during the Weimar Republic. Enabling a form of cultural critique that not only participates in the world created by these new media but also seeks to intervene politically in it points us to the relevance of negative mathematics today.

Who's Afraid of Mathematics? Critical Theory in the Digital Age

At the end of this book, I recall and reformulate an idea posed at its beginning: what would theory look like if it were both critical and mathematical—perhaps even digital? In their theories of history, messianism, and cultural critique, Gershom Scholem, Franz Rosenzweig, and Siegfried Kracauer may not provide a systematic or unified answer to this question. But by returning to the project of negative mathematics that emerged in their work, I hope to show how the set of interdisciplinary tensions found there might shape theory today—theory, that is, that both draws on mathematics and remains rooted in the emancipatory promise of critique. If contemporary scholars working in the digital humanities have already responded to Alan Liu's question ("Where is cultural criticism in the digital humanities?"), then perhaps we should press further and ask: Is critical theory taking full advantage of what digital technologies offer to critical perspectives on culture and art?[1] In what ways do mathematical and computational techniques return us to historical thought rather than obscure it, allowing us to confront the silences and contradictions in the historical record of being and thinking? How can we not only analyze the effects on language and society of the highly advanced mathematical processes found in digital technologies but also employ these same processes more effectively as a means of intervening in society, perhaps through language itself? How can humanists change and reshape the seemingly necessary, but by no means inevitable, categories of "the digital," "the quantitative," and "the computational" through which society thinks and in which we think about society? Formulating answers to these questions, I submit, points to a vision of critical theory—indeed, of the humanities—that would not feel it necessary to emphasize the fact that it, too, is digital. Rather, it would assume that mathematical, quantitative, and digital techniques already existed as part of its epistemological mission and critical repertoire. This would be a critical theory similar to the one envisioned by Scholem, Rosenzweig, and Kracauer, one that is not afraid of mathematics.

My argument throughout this book has been that mathematics—and, in particular, what I have called negative mathematics—has a contribution to make to critical theories of culture, art, and society. In the work of Scholem, Rosenzweig, and Kracauer, this contribution hinged on mathematical approaches to negativity. In the chaotic decades of the early twentieth century, the way that mathematics dealt with absence, lack, division, and privation helped these German-Jewish thinkers refashion language, history, and cultural theory in hopes of realizing the emancipation and inclusion promised by the Enlightenment—not through positivism, but rather by remaining committed to its tradition of critique. For Scholem, Rosenzweig, and Kracauer, this project produced a poetics that employed the restriction of representation in mathematical logic as a way to signify a history that lies beyond the limits of representation. Through infinitesimal calculus, it proposed a messianism that embedded the messianic moment in the here and now, reconfiguring a theory of knowledge to include not only those truths proved by mathematics but also those verified by the experience of individuals and groups. The project argued that a path to a reasonable society lay in a form of criticism that could employ rationalized, geometrical principles of space in order to promote reflective thought in the audience and, hence, spread enlightenment. As an institution and a practice, critical theory drew on ideas similar to these—and can continue to benefit from them in the present—even as chief agents of the Frankfurt School, Max Horkheimer and Theodor W. Adorno, equated mathematics with the rise of Fascism and the epistemological conditions that enabled the Holocaust. Mathematics became a symbol of instrumental reason and of a restrictive, incomplete form of thought, indifferent to politics, history, and aesthetics. Horkheimer and Adorno's initial formulation of critical theory in the 1930s and 1940s defined itself through this opposition to mathematics, and the continuation of the critical project (by theorists such as Jürgen Habermas) adopted and perpetuated it. And yet mathematics did not disappear from the cultural-critical horizon. On the contrary, it has increasingly come to inform the quantitative and digital fabric of the societies in which we live, work, and think.

For Scholem, Rosenzweig, and Kracauer, the turn to mathematics served as an intellectual response to an acute sense of cultural crisis in several arenas, and it is as a response to crisis that I believe we find the enduring relevance of negative mathematics. For these German-Jewish intellectuals, these crises played out on the stage of world history, exemplified above all by World War I. But such crises were also cultural and epistemological, taking root in the fear that the unique contributions offered by the study of aesthetics, history, and religion (just to name a few) were under threat in a world that prioritized rationalization, secularization, and science. It was Horkheimer who perhaps put it best in "Traditional and Critical Theory," published in 1937, when he remarked: "In recent periods of contemporary society, the so-called humanities [*Geisteswissenschaften*] have had but a fluctuating market value and must try to imitate the more prosperous natural sciences whose practical value is beyond question."[2] The trend that Horkheimer described helped motivate his and Adorno's polemic defense of philosophy, and it continues to resonate with contemporary readers, critical theorists and humanists, especially after the economic crisis that started in 2007. While humanists have been speaking of a "crisis in the humanities" at least since the 1960s (the decade that also brought the republication of *Dialectic of Enlightenment*), such sentiments have reached a zenith recently amidst public and political advocacy of the STEM disciplines (science, technology, engineering, and mathematics).[3] In finding avenues of critical thought in negative mathematics, the intellectual project shared by Scholem, Rosenzweig, and Kracauer offers us, I contend, ways to navigate these newest crises in humanistic thought. Negative mathematics offers us not only alternative origins to the critical project but also modes of analyzing and intervening in the digital world, which employ the analytic advantages of mathematics without losing sight of the political, interpretative, and disciplinary imperatives that constitute critical theories of culture and art.

As the preceding chapters have shown, Scholem, Rosenzweig, and Kracauer's work on negative mathematics anticipated how some humanists reacted to a sense of disciplinary crisis a century later, turning to the quantitative and digital approaches to humanistic questions known as the digital humanities. To be sure, humanists have engaged with the digital for quite some time: from the application of computational techniques to textual analysis (so-called "humanities computing") after World War II to the analyses of the philosophical and aesthetic implications of the digital in new German media theory.[4] But, more recently, the term *digital humanities* has come to signify the use of a wide swath of computer-based tools and methodologies designed to facilitate and enhance humanities research. We think of such tools and methods as "digital" in as much as they rely on local and online computer programs to create digital archives or conduct computational, algorithmic analysis on texts, images, and cultural artifacts.[5] As a "discursive construction," the term *digital humanities* has also come to designate a meta-discourse debating how the digital, according to its proponents, revolutionizes the humanities by opening up avenues of inclusion and social emancipation; for others, the term threatens to instrumentalize humanistic inquiry and eliminate critical reflection.[6] Curiously, the current debates surrounding digital humanities echo the dynamics of mathematics as both a potential ally and enemy of humanistic inquiry that we have seen throughout this book. Here, I examine three contemporary impasses in scholarly discourse over the digital humanities that, I believe, negative mathematics and its later eclipse by Horkheimer and Adorno's confrontation with logical positivism can help us reframe and resolve. Although my examination is by no means exhaustive, it points to ways in which the generative negativity of mathematics in Scholem, Rosenzweig, and Kracauer reveals enduring bridges between the humanities and the STEM fields—mathematics in particular. Indeed, negative mathematics suggests that the critical potential of the digital humanities and the enduring significance of mathematics therein lies in using computational approaches to negativity to expose and confront silences, discontinuities, and modes of societal exclusion that persist in the seemingly all-encompassing digital world.

Consider, for example, one of the most vexing issues in the digital humanities: computational approaches to culture and literature known as cultural analytics, microanalysis, and, in its most recognizable form, distant reading.[7] Such approaches use mathematical (algorithmic and statistical) analyses of large corpora of text to look for stylistic and thematic patterns and trends. As it was first popularized by literary scholar Franco Moretti in 2000, computational analyses counter the tradition of close reading, instead using computational techniques, for example, to model figurative language in forty-five hundred poems and uncover statistical patterns in a corpus of four hundred English-language haikus.[8] For proponents of computational reading, such techniques enable scholars to interpret on a mass scale and realize the potential of mass-digitization projects such as Google Books by analyzing a canon of works that would be impossible to read in full, in the traditional sense of the word, in any individual's lifetime.[9] And yet critics of distant reading take exception to the practice, objecting that "digital tools, no matter how powerful, are themselves incapable of generating significant new ideas about the subject matter of humanistic study," as Adam Kirsch put it in the *New Republic*. As adept as computers may be at recognizing patterns in texts, Kirsch claims that interpreting the results of such computation still requires the literary acumen of the studied mind. He asks, "does the digital component of digital humanities give us new ways to think, or only ways to illustrate what we already know?"[10] Responses to Kirsch have aptly pointed out the omissions and blind spots of his criticism.[11] But his analysis pinpoints a nagging epistemological concern attached to computational modes of reading and cultural analysis, namely whether they contribute to the project of knowledge as a whole or merely rearrange and demonstrate through other means what we already know (or could perhaps easily know) by reading the books themselves.

The debate over computational approaches to culture resonates with a viewpoint that holds that the mathematization of thinking cannot tell us anything essentially new in terms of epistemology (see chapters 1 and 2). Hermann Lotze, for instance, held that the true vocation of philosophy was the formulation of the transcendental laws of thought and criticized mathematical logic for only recapitulating in mathematical symbols ideas, like the law of noncontradiction, which were already part of the philosopher's toolbox (see chapter 2).[12] Likewise, Horkheimer and Adorno, in their criticism of logical positivism, argued that the logical positivists' equation of thought with the analytic and tautological statements of mathematics ultimately produced a static vision of knowledge. As Horkheimer wrote in the 1930s, thought executed in the symbols and according to the operations of mathematics not only shut down the ability to comprehend new forms of social existence, but was also epistemologically bankrupt "since it [was] becoming ever more apparent that one can express in principle absolutely nothing with [this modern logic]."[13] In Kirsch, as in Horkheimer and Adorno, this view holds that both mathematical and computational mechanizations of thought exclude the synthetic moment of the human intellect and cannot produce new or meaningful results. Instead, such methods read and think through the repeated tautologies of calculation and equation. And yet the fact that Kirsch cites as a counterexample to the digital humanities Scholem's *Major Trends in Jewish Mysticism*, which was influenced by his work on mathematical logic (see chapter 2), is telling. It suggests the persistence of a view, developed by Horkheimer and Adorno, that fails to see that the seeming tautological repetitions of mathematics or digital technologies can act as a cultural, aesthetic, and interpretive medium.

Against Kirsch, Scholem's negative mathematics implies that one epistemological advantage of computational analyses may lie in their ability to mediate negativity and allow silence, rupture, and discontinuity to speak as constitutive elements of the historical and literary record. For Scholem, mathematical logic tried to eliminate meaning in language and, thus, revealed a way to configure language (the poetry of lament) to symbolize the inexpressibility of the object that it attempts to represent (the historical privations of the Jewish people). Perhaps, then, the epistemological contribution of such distant readings is the metaphor of distance itself—which, as Moretti put it, constitutes "not an obstacle, but a specific form of knowledge: fewer elements, hence a sharper sense of their overall interconnection."[14] Like mathematical logic for Scholem, computational approaches to corpora that are too expansive to read allows us to tune in and represent certain features of cultural objects by tuning other features out. Take, for example, Matthew Jockers's analysis of 758 works of Irish-American prose literature published over the span of 250 years. In examining their publication data, Jockers's study shows how the thesis of a "lost generation" of Irish-American authors in the early twentieth century excludes the majority of publications of the period, for example, those by women further away from the urban centers of the eastern seaboard.[15] Here, distant reading exposes erasure and offers an important corrective to marginalization in literary history. And yet, if we take the example of Scholem's negative mathematics, then the critical potential of computational approaches to culture also lies in fashioning histories that give voice to the gulf between the ineradicable negativity of history (e.g., of loss, erasure, and marginalization) and our limited modes of understanding and expressing it. By analyzing "fewer elements" of corpora from a distance, computational approaches to culture allow us to zero in on negativity—the silences of assimilation, the suffering of exile, and even the erasures of scholarship—and write histories through this negativity, as Scholem did with Jewish mysticism, as the irreducible feature of experience that it indeed is. While current work by digital humanists such as Alexander Gil, Lauren Klein, and Roopika Risam emphasizes the digital humanists' ability to render silence legible, Scholem's negative mathematics points to ways in which digital humanists can further their efforts to capture and address historical negativity by incorporating mathematical and computational negativity as part of their interpretative method.[16]

Another point in the debate about the digital humanities illuminated by negative mathematics can be found in concerns surrounding criticality in digital humanities work. The exclusion of critical thought (see chapter 1) was for Horkheimer and Adorno's initial conception of critical theory the problem inherent in the logical positivists' proposed reduction of reason to protocol statements and mathematical logic. As Horkheimer argued in "The Latest Attack on Metaphysics"—and as he and Adorno claimed in *Dialectic of Enlightenment*—holding thought to the (alleged) immediacy of the empirically given and mathematical operations excluded philosophy's ability to grasp historical consciousness, perpetuated the status quo, and eliminated thought's ability to intervene in both. Contemporary scholars Daniel Allington, Sarah Brouillette, and David Golumbia make a similar claim in their controversial essay in the *Los Angeles Review of Books*, arguing that the digital humanities are marked by "the relative neglect of critical discourse" and an avoidance of "scholarly endeavor that is overtly critical of existing social relations."[17] Indeed, even Liu, a supporter of digital methods, points out that digital humanists rarely "extend their critique to the full register of society, economics, politics, or culture" in the same way that new media theory, to use Liu's example, blends analysis with political critique.[18] I agree that the critical dimension remains a pillar of humanities scholarship, but it seems that this equation of digital humanities with a lack of critical thought hinges, as with Horkheimer and Adorno, on a limited notion of criticality itself. One could object, for instance, that Liu's article associates the origins of cultural criticism with 1968 and, thus, elides longer histories of debate between technology and cultural criticism—in, for instance, the Weimar Republic. Furthermore, as Risam points out, Allington, Brouillette, and Golumbia's criticism neglects the critical potential of digital humanities to provide forms of socially engaged research and pedagogy in non-Western contexts and beyond the major research university.[19] Perhaps expanding our notion of criticality to include other contexts—beyond the 1960s, beyond Horkheimer and Adorno—would provide an image of critical thought and knowledge more fitting for the digital age.

Negative mathematics, especially in Rosenzweig's thought, sheds light on how we might start envisioning such an expansion of criticality and its relationship to instrumentality. Recall that, for Rosenzweig, the way infinitesimal calculus linked nothingness with finitude represented a powerful yet, ultimately, limited tool to reorient epistemology around the individual subject. As a result, the religious participation of individuals (including Jews) became the arbiter of redemption and the reasonable thoughts and actions of individuals and groups became legible as a form of knowledge based on their commitment to and confirmation of an absolute that stands outside of history. Similar to mathematics for Rosenzweig, digital humanities platforms such as Omeka, which enables users to create online archives, and projects such as Minimal Computing, which develops digital methods to work in underserved conditions, open a similar pathway to knowledge that validates and elevates aspects of life, history, and the world overlooked in the digital age.[20] Addressing the enduring lack of access of many individuals and groups to the means of representation, such tools allow communities, such as urban communities of color, with more oblique relationships to technological and institutional power to create, maintain, and preserve cultural histories and legacies. We can think of these platforms as critical in the sense that they open up concepts of representation and knowledge, that, among other things, allow marginalized groups to "verify" (to use Rosenzweig's term) minoritarian cultural traditions within the public negotiation of knowledge. Indeed, if at points narratives of the digital humanities contain utopian undertones, then Rosenzweig's negative mathematics reminds us that a "messianic" element of critical theory lies in the individual's reasonable actions and interactions with Others that stand in for emancipation otherwise deferred in historical experience.[21] How the tools of negative mathematics helped Rosenzweig reframe the idea of knowledge suggests that what digital tools offer is an expansion of epistemology that includes humanistic inquiry and action alongside the standards of proof and utility without relinquishing knowledge's critical relationship to reason.

One final issue surrounding the digital humanities, in which we find echoes of Horkheimer and Adorno's criticism of mathematics, pertains to the political implications of the digital humanities, if not digital technologies in general. Narratives of the digital humanities have often extolled the revolutionary nature of the digital era, which represents, as the collaboratively authored *Digital_Humanities* (2012) contends, "one of those rare moments of opportunity for the humanities," on par with the invention of moveable type and the Industrial Revolution.[22] The digital humanities, the authors continue, thus open up "new modes of scholarship and institutional units for collaborative, transdisciplinary, and computationally engaged research, teaching, and publication."[23] Here the accompanying need for data management, project design, and funding institutions brings digital humanities, as its opponents hasten to point out, into the orbit of the managerial styles of the sciences and business.[24] Critics of the digital humanities claim those types of digital humanities scholarship that remain uncritical of the scientific-industrial origins of their methods only participate in, as Allington, Brouillette, and Golumbia put it, "the neoliberal takeover of the university."[25] According to these critics, an equally suspect politics informs "one of the main reasons why the digitization of archives and the development of software tools . . . can exert such powerful attraction, effectively enabling scholarship to be reconfigured on the model of the tech startup, with public, private and charitable funding in place of Silicon Valley venture capital."[26] Parallel to the digital humanities apparent allergy to critical inquiry, "the fetishization of code and data" does not support emancipation and democracy, but rather "benefits and fits into established structures of institutional power."[27] These authors are correct to demand that we find ways to expose and resist the initiatives that further the corporatization of the academy, in which, to a certain degree, many academics already participate. Acquiescing to the status quo of the technology industry is as reactionary as it is detrimental to (what is left of) the critical autonomy of the humanities and modes of intellectual resistance in the digital age.

And yet the history of mathematics at the origins of critical theory that we have seen unfold in this book cautions us not to conflate a problematic subset of digital humanities work with the implementation of digital technologies in the service of humanistic questions as a whole. Recall that Horkheimer and Adorno made a similar political claim about the logical positivists' turn to mathematics, which the critical theorists likened to liberal politics and economics.[28] Their criticism of mathematics asserted that the ability of the subject to intervene in the world disappears if thought is limited to what is given and its mathematical recurrence. Indeed, Horkheimer's public criticism of the Vienna Circle argued that formalistic epistemology and the division of knowledge into individual branches of the sciences mirrored and reinforced "the prevailing objectives of industrial society with its extremely dubious future."[29] Logical positivism's adherence to the status quo, Horkheimer and Adorno concluded, sanctioned Fascism. In this regard, the political critiques of not only the digital humanities but also digital technologies reiterate Horkheimer and Adorno's interpretive framework. "Within the structures of digital technology," we discern a new "dialectic of enlightenment," according to digital humanists such as David Berry.[30] In the era of Facebook and Google, it is undeniable that, even as digital technologies promise democratization and broader access to knowledge, they also entail further surveillance and new forms of exploitation. But just because such digital giants are involved in a new politics of domination does not necessarily mean that digital technology as such—the conversion of concepts into binary digits and their manipulation through computing machines—is intrinsically or exclusively a mechanism of social control.

I worry that if critical theorists accept such a totalizing intellectual narrative, then we will again subsume mathematics and digital technologies wholesale into their use by corporations, governments, and academic institutions in the service of profit and power. This viewpoint forecloses the notion that critical thought could potentially help shape such technologies and, as Habermas's revision to critical theory reminds us, that instrumental reason constitutes a legitimate and necessary form of human action. If we (again) equate mathematics, computation, and digital technologies with "the project of instrumental reason," then we risk a form of essentialism (mathematics *is* instrumental reason) that not only is false but also relinquishes in full the potential of mathematics as an instrument to help us disrupt and intervene in contemporary structures of power.[31] This is the cautionary lesson of negative mathematics and its eclipse by Horkheimer and Adorno's vision of critical theory.

In contrast to the embrace of business, on the one hand, and the full-scale rejection of mathematics, on the other hand, the project of negative mathematics shared by Scholem, Rosenzweig, and Kracauer holds hope for a more critical constellation of digital technologies, politics, and humanistic inquiry. As one final example, take the image of cultural critique and the role of the cultural critic that emerged in Kracauer's work. For Kracauer, the marginalized figure of the cultural critic, like the Jew in Germany, occupied a vantage point that allowed the critic both to interact with and survey the logic of the materials of mass culture.[32] This dialectical insider-outsider standpoint allowed the critic to recognize and name the incongruences, ruptures, and contradictions that constituted capitalist rationality. To confront and overcome contradiction, as Kracauer wrote in "The Mass Ornament," we must not "flee from the reality" of an increasingly quantified and digital world, but rather work and think "through" it by bridging the materiality of being with the necessity of reason.[33] Integrating critique into the digital humanities is about working "through" digital technologies. "It is not only about shifting the focus of projects so that they feature marginalized communities more prominently," as digital humanist Miriam Posner claims, "it is about ripping apart and rebuilding the machinery of the archive and database so that it does not reproduce the logic that got us here in the first place."[34] Likewise, Kracauer's negative mathematics suggests that the critical potential of the digital humanities lies in their ability to intervene in the technological categories that govern the digital world and to reconstruct these categories on the basis of reason rather than capital. But it also suggests that the critical potential of the digital humanities lies in their ability not only to correct for negativity—for the faulty "logic" of capitalism—but also to expose the enduring contradictions between the logic of society and material modes of existence and represent this negativity in language and history. Instead of relinquishing mathematics, computation, and digital technologies (again) to industry and governance, I contend that the digital humanities can follow the examples of Scholem, Rosenzweig, and Kracauer in utilizing mathematical and computational approaches to negativity to codetermine a digital future that works to grasp and include ideas, voices, and experiences marginalized by the totalizing narratives of big science and industry.

Theory that is critical and mathematical—and, perhaps, even digital—does not require that we abandon the hermeneutic, historical, and critical bedrock of the humanities. On the contrary, theory that borrows from mathematics and, as I hope to have shown in this book, from negative mathematics in particular implies that the significance of the digital humanities lies less in the mathematization of literature than in using mathematical and computational approaches to negativity to reveal the silences, discontinuities, and moments of unrepresentability that persist in the digital age. It entails fashioning, to borrow a phrase from Jockers, a "complementary" vision of technology and theories of art and culture that not only employ but also pay attention to the metaphors through which digital technologies and mathematical processes mediate social and cultural knowledge.[35] It indicates, as media theorist Wendy Hui Kyong Chun put it, that the "dark side" of the digital humanities is also their "bright side," in as much as the digital humanities dialectically return us to questions fundamental to the humanities, such as those of aesthetics and capital.[36] The image of critical thought envisioned by the project shared by Scholem, Rosenzweig, and Kracauer thus compels humanists in the twenty-first century to reincorporate a broader palette of interpretative and analytic tools—whether these be mathematical, quantitative, or digital—as modes of social critique and intervention. At the same time, if we look to Rosenzweig's *The Star of Redemption* or Scholem's *Major Trends in Jewish Mysticism*, we are reminded that integrating mathematics into the humanities does not mean that the much-debated future of critical theory will be exclusively mathematical or digital.[37] Rather, it means that we will be better equipped to analyze and intervene in a future in which power increasingly derives from quantitative, computational, and digital technologies, but only if we do not foreclose them prematurely based on the intellectual rivalries of the past.

ACKNOWLEDGMENTS

I have many people and institutions to thank for their inspiration, encourage-
ment, and generosity since I embarked on this project, around 2005, when I first
caught glimpse of the strange logic of cultures and the surprising (at least then)
historicity of mathematical thought. In terms of institutions, my work has been
supported in innumerable ways by Michigan State University, the College of
Arts and Letters, the Department of Linguistics and Germanic, Slavic, Asian
and African Languages—including two generous grants that allowed this book
to appear in open access format. Special thanks are also due to Gerald Westheimer and the Leo Baeck Institute New York for granting me a Gerald
Westheimer Career Development Fellowship, which enabled archival research
and provided materials essential to the completion of this project. My work
has benefited from the support of the German Academic Exchange Service
(DAAD), which, along with Cornell University's German Cultural Studies
Institute, enabled my participation in the 2014 Faculty Summer Seminar in
German Studies. I could not have written this book without the generosity
of the Deutsches Literaturarchiv (Marbach am Neckar)—as well as the countless archivists and staff in Marbach and at the National Library of Israel who
made my archival research not only possible but also thoroughly enjoyable.
The Deutsches Literaturarchiv, the Udo-Keller-Stiftung, the DAAD, and the
Leo Baeck Fellowship Programme (administered by the Leo Baeck Institute
London and the Studienstiftung des deutschen Volkes) supported early research for this project; it has been a pleasure to continue this work with Programme colleagues through the German Jewish Cultures book series at
Indiana University Press. I would also like to recognize and thank the Deutsches
Literaturarchiv, the National Library of Israel, Suhrkamp Verlag (especially Pe-
tra Hardt), and the University of Michigan Library (Special Collections Research Center) for granting me permissions to reproduce quotations and
images contained in this book.

To the individuals who have supported me and guided my work throughout this project: neither mathematical equations nor words could express my gratitude. The long list of these individuals begins with my *Doktormutter*, Liliane Weissberg, to whom I owe an infinite debt for her skilled advising, ongoing support, and friendship. I have benefited from the guidance of Andrea Albrecht, Paul Fleming, David Kim, Liz Mittman, Pat McConeghy, Paul North, and Benjamin Pollock who tirelessly encouraged my work and deeply shaped my thinking; I thank them for their conversations, attention to detail, and advice. I am grateful for Jerry Singerman's expert guidance in navigating the publication process and for Thomas Lay's adept handling of my manuscript. Thanks are also due to colleagues who read and commented on early versions of this work (those mentioned above as well as Johannes von Moltke) and to scholars and students who generously invited me to their institutions to present snippets of it (Susanna Brogi, Doreen Densky, Marcel Lepper, Peter McIsaac, Kate Schaller, and Kerry Wallach); to these colleagues, I owe a debt of gratitude for our ongoing discussions and—to Anna Kinder, Christian Wiese, Kerry Wallach, and Marcel Lepper—for their advice at conferences, during summers, and over email. A similar debt of support as well as inspiration and patience is also due to my colleagues in the German Program at Michigan State University— Sonja Fritzsche, Senta Goertler, Tom Lovik, Katie McEwen, Liz Mittman, Johanna Schuster-Craig, Lynn Wolff, and Karin Wurst—as well as to members of the MSU community, including David Bering-Porter, Lyn Goeringer, Todd Hedrick, Yelena Kalinsky, Brandon Locke, Kristen Mapes, Ellen McCallum, Jason Merrill, Justus Nieland, Bobby Smiley, Natalie Phillips, Lily Woodruff, and Joshua Yumibe. I would also like to express thanks to Paul Guyer, Gunnar Hindrichs, Eric Jarosinksi, Catriona MacLeod, and Simon Richter for their ongoing intellectual support. My gratitude is also due to the expert editorial advice of Laura Portwood-Stacer. And, finally, I hope to be able to pay back in kind Hannah Eldridge, Peter Erickson, Tatyana Gershkovich, Joela Jacobs, Yosefa Raz, and Sunny Yudkoff for the incisive and constructive feedback that helped shaped the thinking and prose of these chapters. For all those mentioned here, I hope that this book adequately reflects the intellectual enthusiasm and rigor that you have brought to my work throughout the years.

A combination of personal and professional gratitude is due to Osman Balkan, David Bering-Porter, Eric Jarosinski, Jeffery Kirkwood, Mike Laney, Brandon Locke, Kristen Mapes, Pat McConeghy, Bobby Smiley, Maya Vinokour, Leif Weatherby, Caroline Weist, and Lynn Wolff. To Leif, as well as Bobby and David—here's hoping that these collaborations will continue in the future. I owe an eternal debt of gratitude and patience to my parents, Alvin Handelman and Carol Holly, who paved the way for this book, even if it was realized in a rather circuitous—indeed, mathematical—route. And, to Shira Yun, I thank you for your love, support, and much-needed insistence on travel and adventure that have made this book and our lives together, in their many stages and forms, possible.

All translations from German-language sources are my own. Consulted translations are cited in the bibliography.

Chapter 2 is derived in part from the article "After Language as Such: Gershom Scholem, Werner Kraft, and the Question of Mathematics" published in *The Germanic Review*, copyright 2016 by Taylor and Francis, available online at https://doi.org/10.1080/00168890.2016.1190628.

INTRODUCTION

1. See Juola and Ramsay's introduction to mathematics for humanists, *Six Septembers*.

2. Winterhalter, "Morbid Fascination with the Death of the Humanities." For a contrasting view, which upholds the mutual dependence of the humanities and STEM disciplines, see, for example, Koblitz, "Why STEM Majors Need the Humanities."

3. Allington, Brouillette, and Golumbia, "Neoliberal Tools (and Archives)."

4. Although not members of the Frankfurt School, Scholem, Rosenzweig, and Kracauer often appear in histories of critical theory as satellite figures. See, for instance, Jay, *Dialectical Imagination*, 198–199, 21, and 66.

5. As Hitler took power in 1933, the Nazis implemented regulations barring Jews from holding government positions (*Ämter*), which included professorships at the university. He was thus prohibited from lecturing and publishing in Germany, see Smith, *Husserl*, 24–25.

6. Husserl, *Crisis of European Sciences*, 3.

7. These are the words of Husserl's introduction to the speech's publication in *Philosophia* ("Die Krisis der europäischen Wissenschaften und die transzendentale Phänomenologie. Eine Einleitung in die phänomenologische Philosophie," 77). Quoted in *Crisis of European Sciences*, 3.

8. Most locate the origins of the *Sprachkrise* in Hofmannsthal's "Ein Brief," the "Chandos Letter," fictionally addressed to Francis Bacon (1902). See Janik and Toulmin's canonical study, *Wittgenstein's Vienna*, 112–119.

9. Although historical relativism has a deeper conceptual history, the term *crisis of historicism* originates with the publication of Ernst Troeltsch's "The Crisis of Historicism" (1922) and Troeltsch, *Der Historismus und seine Probleme*. In his essay "The Crisis of Science" ("Die Wissenschaftskrisis," 1923),

Kracauer interprets Troeltsch in tandem with Max Weber as contributing attempts to wrest historical analysis from relativism as a means of countering rising discontent with science in contemporary youth culture; see Kracauer, *Werke*, 5.1:596 and 600. See also Beiser, *German Historicist Tradition*, 23–26.

10. Cassirer, *Philosophy of Symbolic Forms*, 3:366–377. Translation modified. On the history of the "foundations crisis" (*Grundlagenkrise*), see Thiel, *Grundlagenkrise und Grundlagenstreit*.

11. The famous example of such results is the "Weierstraß function," which is continuous but not nowhere differentiable. Hahn, "The Crisis in Intuition," 82–83. Klaus Volkert interprets the conflict over intuition as a driving force in the history of modern mathematics, see *Die Krise der Anschauung*.

12. Husserl, *Crisis of European Sciences*, 15.

13. Schmidt-Biggemann, *Topica Universalis*, chaps. 3 and 4.

14. Husserl, *Crisis of European Sciences*, 6.

15. Ibid., 6.

16. See the thesis of Weber's famous 1917 speech; Weber, "Wissenschaft als Beruf," 87 and 111.

17. Husserl, *Crisis of European Sciences*, 23; italics in the original.

18. Ibid., 48–49 and 52.

19. Adorno grew up in a Jewish household in Frankfurt am Main, taking as his pen name his Catholic mother's Corsican family name and retaining the trace of his father's ("Wiesengrund") as the middle initial. Indeed, the relationships among the Frankfurt School, the origins of critical theory, and Jewishness (in particular, a European form of Jewishness) is a topic in need of further exploration, but which ultimately lies beyond the scope of this study. See Jacobs, *Frankfurt School, Jewish Lives, and Antisemitism*; Allen, *End of Progress*.

20. Adorno, *Gesammelte Schriften*, 3:20. Horkheimer and Adorno agree here with Husserl's description of the mathematization of nature in *The Crisis of the European Sciences*, which they cite (ibid., 3:41). The decade before, in their confrontation with logical positivism, Adorno had derided the reification of thought in mathematical logic in comparison to Husserl's "logical absolutism," Adorno and Horkheimer, *Briefe und Briefwechsel*, 4.1:239. See also A 5:48–95.

21. See Hansen, *Cinema and Experience*; von Moltke, *Curious Humanist*.

22. Adorno, *Gesammelte Schriften*, 3:41.

23. Ibid., 3:58.

24. The preface to the first two editions of the text (1944 and 1947) as well as that of the "new edition" (1969) make clear the historical connection that Horkheimer and Adorno have in mind.

25. Rosenzweig, *Gesammelte Schriften*, 3:139.

26. These influences have been canonical since the publication of the first studies on the Frankfurt School, such as Jay's watershed work, *Dialectical Imagination*, chap. 3. Feenberg focuses on the influence of Marx and Lukács; see *Philosophy of Praxis*, chap. 7. See also Susan Buck-Morss's discussion of Adorno and psychoanalysis and his productive disputes with Walter Benjamin in *Origin of Negative Dialectics*, chap. 1 and 9–10. Likewise, Habermas locates in *Dialectic of Enlightenment* a conversation with Nietzsche's notion of the limits of rationality that models a "critique of ideology's totalizing self-overcoming," see Habermas, "Entwinement of Myth and Enlightenment," 107.

27. See the examples in Gamwell, *Mathematics and Art*, x.

28. See *Minima Moralia* (Adorno, *Gesammelte Schriften*, 4:283) and *Dialectic of Enlightenment* (ibid., 3:43), and Hegel, *Phenomenology of Spirit*, 19 and 50–51. On the Frankfurt School's relationship to Hegel and the concept of determinate negation, see Jay, *Dialectical Imagination*, chap. 2; O'Connor, *Adorno*, 48–49; and Cook, *Adorno on Nature*, 83–84.

29. Rabinbach, *In the Shadow of Catastrophe*, 27. Part of the "new Jewish ethos" of messianism as examined by Rabinbach was conditioned by the rare "opportunity," as Paul North calls it, that arose in Central Europe as the meaning of the category "Jew" began to erode; North explores this opportunity in the work of Franz Kafka, a figure much discussed among the thinkers investigated here, North, *Yield*, 1–2. The sense of freedom in chaos runs throughout the historiography of the interwar period; see Weitz, *Weimar Germany*, chaps. 3 and 5.

30. See the A preface for Kant's definition of "critique" (*Kritik*), Kant, *Critique of Pure Reason*, Ax–Axii.

31. Horkheimer, *Gesammelte Schriften*, 4:219.

32. Thanks to the work of Peter Fenves, we have started to take account of mathematics in the work of Walter Benjamin; see Fenves, *Arresting Language*, chap. 6 and Fenves, *Messianic Reduction*, chaps. 4 and 5. In accounts of the history of critical theory focused on Horkheimer and Adorno, mathematics often appears in reference to their criticism of logical positivism and Edmund Husserl's phenomenology; see Jay, *Dialectical Imagination*, 62; Wheatland, *Frankfurt School in Exile*, 104; and Buck-Morss, *Origin of Negative Dialectics*, 105. It also surfaces as a symbol of "formal rational disciplines" next to the natural sciences in discussions of reification and instrumental reason, to which I return in chapter 1; see Feenberg, *Philosophy of Praxis*, 78–79.

33. Mendelssohn, "On the Evidence in Metaphysical Sciences," 258.

34. Ibid., 255.

35. Altmann, *Moses Mendelssohn*, 116.

36. Kant, *Critique of Pure Reason*, Bxxii. On Kant's "Copernican Revolution" see Guyer, *Kant*, chap. 2. Kant makes the latter argument in the "Transcendental Aesthetic," *Critique of Pure Reason*, A19/B33–A48/B73.

37. Maimon was responding to the analysis of infinitesimal calculus (and the concept of the infinite) written by Lazarus Bendavid (a German Jew and Maimon's intermittent patron) and published in 1789 as *Versuch einer logischen Auseinandersetzung des Mathematischen Unendlichen*; *Essay on Transcendental Philosophy*, 152–153.

38. Maimon, *Essay on Transcendental Philosophy*, 21. For an explanation of Maimon's use of differentials, see Duffy, "Maimon's Theory of Differentials as the Elements of Intuitions."

39. Or so Maimon tells us in his autobiography. See especially the first three chapters of Maimon, *Solomon Maimon*, 6–31 and 275. On Maimon's acculutration, science, and his peers, respectively, see Weissberg, "Erfahrungsseelenkunde als Akkulturation: Philosophie, Wissenschaft und Lebensgeschichte bei Salomon Maimon," 327; and Weissberg, "Toleranzidee und Emanzipationsdebatte: Moses Mendelssohn, Salomon Maimon, Lazarus Bendavid," 376–378.

40. See Goetschel, *Discipline of Philosophy and the Invention of Modern Jewish Thought*, 60–66.

41. Although Cohen did not remark on this affinity with Maimon, his contemporaries did; see Kuntze, *Die Philosophie Salomon Maimons*, 339; and Bergmann, "Maimon und Cohen," 548. On Cohen's reception of Maimon, see Poma, *Critical Philosophy of Hermann Cohen*, 279n10; and Albrecht, "'[H]eute gerade nicht mehr aktuell,'" 54–55.

42. Cohen works this out in Cohen, *Werke*, 5:14 and 27–41.

43. Ibid., 6:33.

44. Ibid., 6:84–89. Cohen draws on Kant's notion of the infinite judgment in *Critique of Pure Reason*, A72–73/B97–98.

45. Cohen, *Die Religion der Vernunft aus den Quellen des Judentums*, 73.

46. This tendency is perhaps most prominent in regard to Rosenzweig's *Star of Redemption*; as Amos Funkenstein writes: "Rosenzweig did not make up his mind whether the 'formal' language (symbols) with which" he speaks about the ideas of God, World, and Self, "*is* mathematics or is only *like* mathematics. But of course it could only be the latter." See Funkenstein, *Perceptions of Jewish History*, 283 and 289. Peter Gordon makes a similar point in "Science, Finitude, and Infinity," 39–51. For an insightful and contrasting view, see Smith, "Infinitesimal as Theological Principle," 563. I will return to this debate in chapter 3.

47. Blumenberg, *Paradigms for a Metaphorology*, 5.

48. Ibid., 20–28.

49. On the experience of German-Jewish intellectuals and World War I, see Brenner, *Renaissance of Jewish Culture in Weimar Germany*, 31–35; and Mosès, *Angel of History*, 11.

50. As in Scholem's 1964 essay "Against the Myth of the German-Jewish Dialogue." See also Hess, *Germans, Jews and the Claims of Modernity*, 10–11.

51. Theology entered the secondary literature through Benjamin's theses "On the Concept of History" (Benjamin, *Gesammelte Schriften*, 1:694) and Adorno's *Minima Moralia*; see Jay, *Dialectical Imagination*, 199–201, 277–279; and Buck-Morss, *Origin of Negative Dialectics*, 94–95, 168–173. On the origins of the theological-messianic impulse in Benjamin, see Rabinbach, *In the Shadow of Catastrophe*, chap. 1; and Tiedemann, "Historical Materialism or Political Messianism?"

52. On the link between aesthetics and redemption in Adorno and Benjamin, see Buck-Morss, *Origin of Negative Dialectics*, chap. 8; Brunkhorst, *Adorno and Critical Theory*, chap. 4; Wolin, *Walter Benjamin: An Aesthetic of Redemption*; Wellmer, "Truth, Semblance, Reconciliation"; and Hohendahl, *Fleeting Promise of Art*.

53. See, for instance, Peter Gordon's analysis of Adorno's "inverse theology" in *Adorno and Existence*, 173–182.

54. Klein, "Image of Absence." See also Risam, "Beyond the Margins."

55. Pollock, *Franz Rosenzweig*.

1. THE TROUBLE WITH LOGICAL POSITIVISM

1. For the full breadth of the dispute between critical theory and logical positivism in both the 1930s and the 1960s, see Hans-Joachim Dahms, *Positivismusstreit*; O'Neill and Uebel, "Horkheimer and Neurath." The dispute has been part of the history and internal debates of the Frankfurt School since the late 1960s; see Wellmer, *Critical Theory of Society*, 9–18; and Jay, *Dialectical Imagination*, 61–63.

2. Horkheimer's letter to Adorno from February 22, 1937, leaves little doubt that academic resources were part of his criticism: "In the end, this magic," namely, logical positivism, "is all about academic positions and endowed professorships [*ordentliche Lehrstühle*]," *Briefe und Briefwechsel*, 4.1:294. The members of the Vienna Circle, Dahms explains, had no better chances in exile than the Frankfurt School, Dahms, *Positivismusstreit*, 141–142.

3. Adorno, *Gesammelte Schriften*, 3:42 (hereafter cited in text as A, followed by volume and page number).

4. See, for instance, Jay, *Dialectical Imagination*, 61–63.

5. See Adorno's letter to Benjamin dated July 2, 1937, in Adorno and Benjamin, *Briefe und Briefwechsel*, 1:258. Benjamin and Adorno's official report composed an official report for the Institute, "Kongreß für Einheit der Wissenschaft (Logische Positivisten)" in Adorno and Horkheimer, *Briefe und Briefwechsel*, 4.1:560–570.

6. Horkheimer had been in conversation especially with Neurath over a planned collaboration; on this fascinating prehistory to the debate, see Dahms, *Positivismusstreit*; O'Neill and Uebel, "Horkheimer and Neurath."

7. Adorno and Horkheimer, *Briefe und Briefwechsel*, 4.1:194.

8. Carnap, "Die alte und die neue Logik," 16; Neurath, "Wissenschaftliche Weltauffassung," 306. I cite from the English translation contained in Neurath's *Empiricism and Sociology* (1973), see also Hahn, Neurath, and Carnap, *Wissenschaftliche Weltauffassung*. The authors even cite Vienna as fertile political "ground" for their worldview: "In the second half of the nineteenth century, liberalism was long the dominant political current. Its world of ideas stems from the enlightenment [*sic*], from empiricism, utilitarianism and the free trade movement of England," 301.

9. The logical positivists had just as much reason to fear the rise of Fascism as the Frankfurt School. On the one hand, the Vienna Circle as well as logical positivism was coded in interwar Vienna as "Jewish," see Silverman, *Becoming Austrians*, 60–64. On the other hand, as noted by A. J. Ayer as well as Dahms, members of the circle such as Neurath were politically active on the Left, and Neurath was convicted in the Bavarian revolutions of 1919; see Dahms, *Positivismusstreit*, 38; Ayer, "Editor's Introduction," 7. On their paths of exile, see Dahms, "Emigration of the Vienna Circle."

10. Both Horkheimer and Adorno completed their dissertations with the Neo-Kantian, Hans Cornelius, while Carnap completed his with Bruno Bauch. See Dahms, *Positivismusstreit*, 22–28. Peter Gordon provides a helpful discussion of Neo-Kantianism in the early twentieth century in *Continental Divide*, 43–86.

11. Carnap, for instance, provides a particularly polemical stance against philosophy in his introduction to Carnap, *Unity of Science*, 21–30. This is the English translation of Carnap, "Die physikalische Sprache als Universalsprache der Wissenschaft." On the historical development and continuing significance of logical positivism, see Friedman, *Reconsidering Logical Positivism*, pt. 3.

12. Neurath, "Wissenschaftliche Weltauffassung," 308.

13. Ibid., 307.

14. Horkheimer, *Gesammelte Schriften*, 4:112 (hereafter cited in text as H, followed by volume and page number).

15. Instrumental reason forms one of the canonical charges against mathematics in the historiography of critical theory. "The critique of the positivist understanding of science," Habermas writes of *Dialectic of Enlightenment*, is heightened "to the totalized reproach that the sciences themselves have been absorbed by instrumental reason" ("Entwinement of Myth and Enlightenment," 111). Thomas Wheatland sees Horkheimer's confrontation with the Vienna Circle as anticipatory of the later critique of "instrumental rationality" (*Frankfurt School in Exile*, 118).

16. Andrew Feenberg, for instance, argues that the Frankfurt School draws on Lukács's *History and Class Consciousness* (and Emil Lask) in interpreting mathematics (and the natural sciences) in terms of the reification of intellectual forms. For Lukács, mathematics, as Feenberg puts it, fails to take account of the "purely contingent or 'factical' [*sic*] objects to which they refer" (*Philosophy of Praxis*, 78 and 102–103). Parallel to reification, mathematical calculation (*Berechenbarkeit*) enters critical theory, as in Hauke Brunkhorst's intellectual history of Adorno, through Weber's thesis on "domination through calculation" (*Adorno and Critical Theory*, 41); Dubiel, *Theory and Politics*, 91. Indeed, Adorno extends the criticism of mathematics to the "reification of logic," which runs parallel to the commodity form, in Martin Jay's words, in his work on Husserl, see Jay, *Dialectical Imagination*, 69. See also Buck-Morss, *Origin of Negative Dialectics*, 11; and Wiggershaus, *Frankfurt School*, 533.

17. Carnap, "Die alte und die neue Logik," 12–14. On mathematical logic, see Peckhaus, *Logik, Mathesis Universalis und allgemeine Wissenschaft*; "The Mathematical Origins of Nineteenth-Century Algebra of Logic"; and Grattan-Guinness "The Mathematical Turns in Logic."

18. Benjamin, *Gesammelte Schriften*, 2.1:143.

19. Kant, *Critique of Pure Reason*, B10–B24; example on B16.

20. Neurath, "Wissenschaftliche Weltauffassung," 308 and 311.

21. Carnap, *Unity of Science*, 33.

22. Wittgenstein, *Tractatus Logico-Philosophicus*, 3.

23. Adorno and Horkheimer, *Briefe und Briefwechsel*, 4.1:206.

24. Mathematical logicians from Leibniz to Wittgenstein, for instance, form one of the groups of thinkers that the logical positivists cite as their intellectual influences in *The Scientific Conception of the World* (along with Feuerbach and Marx in the category of "hedonism and positivist sociology"); Neurath, "Wissenschaftliche Weltauffassung," 304 and 309–310. The interest in mathematical logic stems from Carnap; see his 1929 *Abriss der Logistik*.

25. The intensity and circulation of Scholem and Benjamin's work on mathematics bears noting. The two became friends around July 1915, in part,

over discussion of mathematics, see Scholem, *Tagebücher*, 1:134. According to Tiedemann and Schweppenhäuser, Benjamin's return to the subject of mathematics and Russell's paradox occured between the middle of 1916 and the middle of 1917 (Benjamin, *Gesammelte Schriften*, 6:639–640). That November, Scholem held his *Referat* on mathematical logic and, subsequently, Benjamin asked Scholem repeatedly to see the *Referat*—as far as I can tell, to no avail. See Benjamin, *Gesammelte Briefe*, 1:404, 407, and 418.

26. On the relationship between mathematics and Benjamin's theory of language see Fenves, *Arresting Language*, 115–125. See also Fenves, *Messianic Reduction*, 117 and 122–124.

27. Russell formulated his paradox in set-theoretical terms, the details of which would divert too far from our analysis of mathematics and critical theory. Briefly, Russell's paradox follows from the idea of a class with the defining predicate that the members of this class do not predicate themselves. The contradiction arises when we ask if this defining predicate predicates itself and, thus, if it belongs to the class it predicates or not. If it does predicate itself, then it belongs to the class it predicates, meaning, by the definition of the class, that it does not predicate itself, which is a contradiction; if it does not predicate itself (i.e., it is not among the predicates in the class that it predicates), then it does belong to the class it predicates. See Russell's definition in *Principles of Mathematics*, 80; see also Russell's discussion in ibid., chap. 10. For a more detailed account of Benjamin's encounter with Russell's paradox and its consequences, see Fenves, *Messianic Reduction*, 125–130.

28. Benjamin, *Gesammelte Schriften*, 6:9.

29. I will return to this essay in chapter 2. See also Buck-Morss, *Origin of Negative Dialectics*, 88–90.

30. Benjamin, *Gesammelte Schriften*, 6:9–10.

31. Fenves, *Messianic Reduction*, 116. See also Benjamin, *Gesammelte Schriften*, 2.2:601–602.

32. Benjamin, *Gesammelte Schriften*, 1:207.

33. Adorno's lecture borrows at multiple points from Benjamin's *Origin of the German Tragic Drama*, not only as we see in terms of mathematics but also regarding the notion of the "intentionless" nature of reality; see Eiland and Jennings, *Walter Benjamin*, 359; and Fenves, *Messianic Reduction*, 1.

34. This text was Adorno's *Antrittsvorlesung*, the public lecture given when Adorno received the *venia legendi* at the University of Frankfurt. In it, Adorno contextualizes his philosophical proposal amidst contemporary reactions to the realization of philosophy's inability at comprehending the totality of the real after idealism. On Heidegger, see A 6.

35. Adorno and Horkheimer, *Briefe und Briefwechsel*, 4.1:239–242. Adorno's letter makes these points in reference to Russell's paradox, which, as Dahms notes, he likely took from Benjamin; see Dahms, *Positivismusstreit*, 89n218. Adorno and Benjamin's proposed solutions are similar: both claim the paradox arises when Russell assigns "meaning" to a meaningless sentence like "this sentence is false," instead of recognizing it as "complexes fixed in words and sounds" (Benjamin, *Gesammelte Schriften*, 6:10) or "a mere complex of words" (Adorno and Horkheimer, *Briefe und Briefwechsel*, 4.1:241). Adorno's likely adaptation does away with Benjamin's terminology, such as predicable, focusing more on the notion of the "word" as a unit of meaning rather than "sign" (*Zeichen*).

36. Adorno and Horkheimer, *Briefe und Briefwechsel*, 4.1:241–242. Lukács makes a similar point in his first "antinomy of bourgeois thought"; see Lukács, *History and Class Consciousness*, 120.

37. Adorno and Horkheimer, *Briefe und Briefwechsel*, 4.1:253.

38. Ibid., 4.1:294.

39. Ibid., 4.1:279. As Wiggershaus claims, the idea of "bringing" something "to language" ("etwas zum Sprechen bringen") is a leitmotif throughout Adorno's philosophy; see Wiggershaus, *Wittgenstein und Adorno*, 123.

40. On the importance of "determinate negation" ("bestimmte Negation") for Adorno—which *Dialectic of Enlightenment* refers to here as "determinative negation" ("bestimmende Negation")—and its origins in Hegel, see O'Connor, *Adorno*, 48–49; and Cook, *Adorno on Nature*, 83–84. See also note 28 in this book's introduction.

41. See A 11:16–22, here 18–19.

42. See the contributions of Eva Geulen and Andrew Hewitt in Fisher and Hohendahl, eds., *Critical Theory: Current State and Future Prospects*.

43. Buck-Morss, *Origin of Negative Dialectics*, 13.

44. The Odysseus scene has become a standard interpretation of *Dialectic of Enlightenment*; see Jay, *Dialectical Imagination*, 263–266.

45. Adorno and Horkheimer, *Briefe und Briefwechsel*, 4.1:244.

46. Carnap, *Unity of Science*, 32. Both Wittgenstein and Ernst Mach developed similar ideas of basic, scientific statements; Wittgenstein calls his "elementary proposition" (*Elementarsatz*). Wittgenstein, *Tractatus Logico-Philosophicus*, 42. Mach's were part of his "atomistic positivism," as Carnap calls it (47).

47. Carnap, *Unity of Science*, 43–44. On "the elimination of experience," see Hanfling, "Logical Positivism."

48. Carnap, "Über Protokollsätze," 224.

49. Carnap, *Unity of Science*, 32. The logical positivists, especially Neurath and Carnap, debated the interpretation of the concept and function of protocol sentences (as Uebel shows), the nuance of which is lost in Horkheimer and Adorno's interpretation. See Uebel, *Empiricism at the Crossroads*; Carnap, "Über Protokollsätze"; and Neurath, "Protokollsätze."

50. Neurath, "Wissenschaftliche Weltauffassung," 308–309; translation modified.

51. See, for instance, Jay, *Dialectical Imagination*, 54 and chaps. 3 and 6.

52. Adorno and Horkheimer, *Briefe und Briefwechsel*, 4.1:242–243.

53. Ibid., 4.1:242.

54. Wittgenstein, *Tractatus Logico-Philosophicus*, 5.

55. On the concept of second nature in Lukács and Adorno, see Buck-Morss, *Origin of Negative Dialectics*, 55–57.

56. Adorno and Horkheimer, *Briefe und Briefwechsel*, 4.1:255.

57. Horkheimer's phrase "der Rest ist Schweigen" matches August Wilhelm Schegel's translation of *Hamlet*, first published in 1798 as William Shakespeare, *Shakspeare's dramatische Werke*, 361. I would like to thank Jocelyn Holland for pointing out this connection.

58. On the history of the logical positivists' relationship to Nazism, see Beller, *Vienna and the Jews, 1867–1938*, 16. Allegedly, Horkheimer separated the "laudable intentions of individual Positivists" from the "objectively reactionary function of their philosophy," see Dahms, *Positivismusstreit*, 147.

59. See, for instance, discussions by later critical theorists, such as in Habermas, *Theory of Communicative Action*, 1:366–399; Benhabib, *Critique, Norm, and Utopia*, chap. 5.

60. Adorno and Horkheimer, *Briefe und Briefwechsel*, 4.1:196.

61. Here I cite from Horkheimer, *Eclipse of Reason*, 7, 14, and 16. On the expansion of Horkheimer's critique to include American pragmatism, see Wheatland, *Frankfurt School in Exile*, chap. 3.

62. Horkheimer, *Eclipse of Reason*, 16. Horkheimer later argues, "positivism is philosophical technocracy. It specifies as the prerequisite for membership in the councils of society an exclusive faith in mathematics" (41). A more detailed examination of the relationship between mathematics and technology to the notions of technocracy and "the administered world" is beyond the scope of this study (A 3:10).

63. Horkheimer, *Eclipse of Reason*, 12.

64. Feenberg, *Philosophy of Praxis*, 216. Feenberg has taken major steps toward reincorporating technology into the critical project, arguing that the reactionary political stance associated with technologies lies less "in technol-

ogy per se but in the antidemocratic values that govern technological development" (*Critical Theory of Technology*, 3).

65. Neurath, "Wissenschaftliche Weltauffassung," 307.

66. Carnap, "Die alte und die neue Logik," 13.

67. Neurath, "Wissenschaftliche Weltauffassung," 310.

68. Adorno and Horkheimer, *Briefe und Briefwechsel*, 4.1:195.

69. The logical positivists' exclusion of unverifiable and nonmathematical statements from knowledge hit a common nerve for those interested in other fields of study—such as history, language, and society. Horkheimer "must have felt especially attacked," Dahms thinks, "when the Positivists seemed to have dismissed his theory of a better society as not only false but also—as suspected metaphysics—senseless" (*Positivismusstreit*, 139).

70. Lukács, *History and Class Consciousness*, 110–149.

71. Ibid., 117; see also Feenberg, *Philosophy of Praxis*, 79.

72. Lukács, *History and Class Consciousness*, 117.

73. Adorno and Horkheimer, *Briefe und Briefwechsel*, 4.1:239.

74. Ibid., 4.1:294.

75. Lukács points as well to "the rejection of every 'metaphysics'" in the "antinomies of bourgeois thought" as a "renunciation" explicit in the work of Ernst Mach and Henri Poincaré, among others. Lukács, *History and Class Consciousness*, 120.

76. Neurath, "Wissenschaftliche Weltauffassung," 310.

77. H 4:110 and 11; see also "Traditional Theory and Critical Theory," in H 4:176–178, here 177.

78. See Habermas, "Entwinement of Myth and Enlightenment"; and Wellmer, *Critical Theory of Society*, 136.

79. See the title essay in Habermas, "Modernity: An Unfinished Project."

80. Habermas, "Entwinement of Myth and Enlightenment," 111.

81. Wellmer, *Critical Theory of Society*, 16.

82. Ibid., 136 and 138.

83. On the return to the philosophy of science in critical theory, see Honneth, *Pathologies of Reason*, 168–171.

84. See, for instance, Horkheimer, *Critique of Instrumental Reason*.

85. Düttmann, *Philosophy of Exaggeration*, chaps. 1 and 2.

86. On Marcuse's later work in relation to Horkheimer and Adorno, see Feenberg, *Philosophy of Praxis*, ix and 171–174.

87. Marcuse, *One-Dimensional Man*, 143–144. Marcuse was aware of the work of the logical positivists as recorded in his review of the *International Encyclopedia of Unified Science* in the 1930s. The review criticizes the notions of

"reason, freedom, happiness, and tolerance" excluded by logical positivism. Marcuse, "Review of International Encyclopedia of Unified Science," 228.

88. Marcuse, *One-Dimensional Man*, 143.

89. See part 2 of Dahms, *Positivismusstreit*, chaps. 4 and 5.

90. Here the target is, again, "Carnap." Adorno's introduction to the volume also repeats the claim that, "resting on the authority of science," the positivists "wish to liquidate philosophy" (*liquidieren möchen*); recall that Adorno's *Antrittsvorlesung* used the same expression three decades earlier, A 8.1:285 and, in 1931, A 1:331.

91. Honneth, *Pathologies of Reason*, 168; capitalization modified in translation.

92. Habermas, *Knowledge and Human Interests*, 308.

93. Ibid., 68.

94. Ibid., 67–68. Habermas reiterates this point in *Theory of Communicative Action*, 1:375.

95. Although Adorno's first letter to Scholem invokes the concept of the "expressionless" (*das Ausdrucklose*), which is related to his and Benjamin's work on mathematics (see chapter 2), their correspondence focuses in the main on reconstructing the legacy of Benjamin. See Adorno and Scholem, "*Der liebe Gott wohnt im Detail*," *Briefwechsel 1939–1969*, 11. See also Weissberg, *Über Haschisch und Kabbala*.

2. THE PHILOSOPHY OF MATHEMATICS

1. Scholem, *Tagebücher*, 2:67 (hereafter cited in text as S, followed by volume and page number).

2. Weber, *Gesamtausgabe*, 1.17:86–87.

3. On Scholem and exile, see Engel, *Gershom Scholem*, 26–61. On World War I, see also Biale, *Gershom Scholem: Kabbalah and Counter-History*, 16–22; and Mosès, *Angel of History*, 12–13.

4. See Scholem's open letter to Manfred Schlösser, "Against the Myth of the German-Jewish Dialogue," 61–64.

5. On the Enlightenment politics of Jewish emancipation, see Hess, *Germans, Jews and the Claims of Modernity*, especially 28–32. On Judaism in Scholem's household, see Scholem, *Von Berlin nach Jerusalem*, 10–11. As Scholem writes: "One sees that with us nothing is left of a Jewish family. After 75 years!! Hopefully it will be different with me one day" (S 1:11 and 158). Important in this context is also Scholem's position on what he saw as the contradictory nature of the German-Jewish experience as self-deceit; see Scholem, "With Gershom Scholem," 1–7; and "On the Social Psychology of the Jews in Germany: 1900–1933," 18–19.

6. Scholem's criticism of his father was returned in kind; as he recalls his father's disapproval of his interest: "My son the gentleman engages in nothing but unprofitable pursuits. My son the gentleman is interested in mathematics, pure mathematics. I ask my son the gentleman: What do you want? As a Jew, you have no chance of a university career. You cannot get an important position. Become an engineer and go to a technical college, then you can do as much math in your free time as you like. But no, my son the gentleman does not want to become an engineer, he wants only pure mathematics. My son the gentleman is interested in Jewishness [*Jüdischkeit*]. So, I say to my son the gentleman: Please, become a rabbi, then you can have all the Jewishness you want. No, my son the gentleman in no way wants to be a rabbi. Unprofitable pursuits" (*Von Berlin nach Jerusalem*, 71). Scholem's anarchism has been a salient feature of both his critical and popular reception; see Engel, *Gershom Scholem*, 55–61; Biale, *Gershom Scholem*, 5–6; and Prochnik, *Stranger in a Strange Land*, 20.

7. See, for instance, Jay, "Politics of Translation"; Benjamin, *Rosenzweig's Bible*; and Engel, *Gershom Scholem*.

8. See Fenves, *Messianic Reduction*, 103–124. See also Fenves, *Arresting Language*, 174–226.

9. Scholem writes: "Like all their spiritual kin among Christians or Moslems [*sic*], the Jewish mystics cannot, of course, escape from the fact that the relations between mystical contemplation and the basic facts of human life and thought is highly paradoxical. . . . How is it possible to give lingual expression to mystical knowledge, which by its very nature is related to a sphere where speech and expression are excluded? How is it possible to paraphrase adequately in mere words the most intimate act of all, the contact of the individual with the divine?" (*Major Trends in Jewish Mysticism*, 14–15). That redemption (real or in language) is foreclosed by the same conditions that would realize it likewise constitutes the paradox that ends Scholem's essay "The Messianic Idea in Judaism," 34–35.

10. Weber, *Benjamin's -abilities*, 5–6. On the significance of the concept of tradition in Scholem, see Alter, *Necessary Angels*; and Schwebel, "Tradition in Ruins."

11. Scholem, "Zehn unhistorische Sätze," 264. If Scholem's narratives of Jewish mysticism reflected his biography, as Amir Engel shows, then the metaphysics expressed in this passage and afforded by mathematics was also "one of the narratives, indeed stories" into which Scholem "molded the vast literature of Jewish mystical lore" (*Gershom Scholem*, 18).

12. See Mehrten's categorization of mathematics around the turn of the century into a "modern" movement (Hilbert, Zermelo, Cantor) and a

counter-modern movement (Klein, Poincaré, etc.) in *Moderne, Sprache, Mathematik*, 108–186. On the key mathematical-historical studies that discuss the "Grundlagenkrise," see Mehrtens, *Moderne, Sprache, Mathematik*. See also Grattan-Guinness, *Search for Mathematical Roots*; and Thiel, *Grundlagenkrise und Grundlagenstreit*.

13. See Volkert, *Die Krise der Anschauung*, xix–xxii. See also Hahn, "Crisis in Intuition," 74–77.

14. Plato, *Republic*, 522c.

15. Rosenzweig, *Gesammelte Schriften*, 3:392. Meant here is the fifth edition of Nernst and Schönflies, *Einführung in die mathematische Behandlung der Naturwissenschaften*. See chapter 3.

16. Voss, *Über das Wesen der Mathematik*, 2. Nernst and Schönflies write in *Einführung in die mathematische Behandlung der Naturwissenschaften*, v: "In general, one can say that a natural-scientific discipline turns ever more frequently to the methods of higher mathematics for the expansion and deepening of results won from direct observation, the more further advancements are made in the theoretical handling of immediate experimental results."

17. As Scholem concisely puts it in early 1917, a divine council had been held in his honor to determine his fate, spurning him onto intellectual greatness: "Because of this, the concept of the science, which alone deserves to be named the introduction to the Torah, was revealed to me, the introduction to the teaching of order or the teaching of the spiritual order of things" (S 1:468. Although there are many potential translations of "die Lehre" (e.g., "doctrine," "instruction"), I chose "the teachings" in order to retain its semantic proximity to the Hebrew word *Torah*.

18. Kracauer, *Werke*, 5.1:591–601. See Glatzer, who writes of Rosenzweig: "A Western European intellectual, he was a proud heir of the nineteenth century. This was the bourgeois world of faith in progress, a faith assured by the steady development of science; the evolution of man and of society appeared inevitable" (*Franz Rosenzweig*, xi). Cf. Mendes-Flohr and Reinharz, "From Relativism to Religious Faith." Recently, Pollock has challenged the long-held belief that Rosenzweig underwent a conversion from a scientific to a religious world view during the course of his famed 1913 *Leipziger Nachtgespräch* with Eugen Rosenstock by emphasizing Rosenzweig's early encounters with the philosophy of Marcion of Sinope; see Pollock, *Franz Rosenzweig's Conversions*, especially 47–50.

19. See standard interpretations of the so-called "Sprachkrise" in Janik and Toulmin, *Wittgenstein's Vienna*.

20. Lazier, *God Interrupted*, 151.

21. As Scholem notes in his autobiography, by the time he arrived in Munich, he was planning to write a dissertation on the "linguistic philosophy of the Kabbalah," see Scholem, *Von Berlin nach Jerusalem*, 141. In the 1960s, he completed the project as "The Name of God and the Linguistic Theory of the Kabbalah"; see *Judaica* 3:7–70. See also Wolosky, "Gershom Scholem's Linguistic Theory," 165; and Weidner, *Gershom Scholem*, 174–196.

22. On November 11, 1916, Benjamin writes to Scholem: "a week ago, I started a letter to you that encompassed eighteen pages. It was an attempt to answer in context the not small number of questions that you put before me." In his memoirs, Scholem claims he first raised questions "about the relationship between mathematics and language" in a letter to Benjamin before November, 1916; see *Walter Benjamin*, 48. Scholem's correspondence with Werner Kraft reveals that before departing to Switzerland, Benjamin entrusted his draft response to Scholem along with other manuscripts and letters to Kraft, which the latter reports having read in an unpublished letter; see Kraft to Scholem, July 18, 1917, Gershom Scholem Archive, National Library of Israel. Both Scholem's pre-November 1916 letter to Benjamin on mathematics and language and Benjamin's eighteen-page response have been lost. Benjamin's collected works contain notes on the relationship between language and mathematics, entitled "Notes continuing the work on language." Benjamin, *Gesammelte Schriften*, 7:785–790. See also Fenves, *Messianic Reduction*, 277n13.

23. See interpretations in Fenves, *Arresting Language*, 201–206. For an interpretation of this essay in the context of Benjamin's work on language essay, see Menninghaus, *Walter Benjamins Theorie der Sprachmagie*, 9–49.

24. Benjamin, *Gesammelte Schriften*, 2.1:143.

25. Ibid., 6:11. Compare with Fenves's interpretation, which links Benjamin's concepts of "proper" and "improper" meaning to his theories of language and time; see Fenves, *Messianic Reduction*, 125–130.

26. The terms *structure* and *construction* (as in *Aufbau*) permeate the work of the logical positivists; see, for example, Carnap's *Habilitation* thesis, *The Logical Structure of the World*. As Peter Galison shows, metaphors of construction facilitated collaborations between the positivists and the Bauhaus; see Galison, "Aufbau/Bauhaus."

27. Poincaré, *Wissenschaft und Hypothese*, 1. On the sociology of mathematics as a "discipline of proof," see Heintz, *Die Innenwelt der Mathematik*.

28. Guyer, *Kant*, 47.

29. Kant, *Critique of Pure Reason*, B19–21 and A38–39/B55.

30. Ibid., B15.

31. For the legacy of these issues, see Friedman, *Parting of the Ways*.

32. See the works collected in Benacerraf and Putnam, *Philosophy of Mathematics*.

33. Scholem's work on Novalis deserves closer attention beyond what can be paid in this study. He first comes into contact with Novalis's aphorisms on mathematics in September 1915 (S 1:153) in an 1837 edition, published by Ludwig Tieck and Friedrich Schlegel, of *Novalis Schriften*, 2:145–149. Yet Scholem's usage of the phrase "Mathematik der Mathematik" is curious, because it does not appear in this specific edition, while it appears in later collections of Novalis's fragments. Indeed, it is a crucial part of Novalis combination of mathematics and poetry based on the idea of exponentiation (*potenzieren*); see Novalis, *Schriften*, 3:245 and 168. For more on Novalis's use of mathematics, see Bomski, *Die Mathematik im Denken und Dichten von Novalis*.

34. Poincaré, *Wissenschaft und Hypothese*, 13.

35. Ibid., 9, 17 and 50.

36. At first sight, it is tempting to jump at the potential similarities between mathematics, as proposed here and Kabbalah; according to one of Scholem's more contentious contemporaries, Oskar Goldberg, the Pentateuch can be understood as a "system of numbers" unfolding out of basic numbers (*Grundzahlen*) corresponding to the letters of the name of God.

37. An unpublished diary entry from August 1, 1915 (omitted in S 1:140) expands Scholem's reflections on the synthetic-analytic nature of mathematical judgments. In the omitted portion of the entry, as Julia Ng describes in detail, Scholem definitively rejects the idea that mathematics contains synthetic judgments by examining the cases of definitions, axioms, and postulates, see Ng, "'+1': Scholem and the Paradoxes of the Infinite," 201–202. See Gershom Scholem's Diaries, 29. For Hilbert, for instance, axioms were the "definitions [of the basic concepts of a scientific system]," see Voss, *Über das Wesen der Mathematik*, 106–107.

38. Scholem to Lissauer, October 20, 1916, in Gershom Scholem Archive, 2.

39. Mauthner, *Beiträge zu einer Kritik der Sprache*, 3:295–297.

40. Kant, *Critique of Pure Reason*, A714/B742.

41. Ibid., A727/B755 and A730/B758.

42. Ibid., A731/B759.

43. Diary entry by Scholem, August 1, 1915, in Gershom Scholem's Diaries, 29. In the original: "eine willkürliche Namengebung. Das Ding, das nur einmal zwischen 2 Punkten da ist, nennen wir—gleich ob es existiert oder nicht—wir nennen es erst einmal gerade Linie."

44. Russell and Whitehead, *Principia Mathematica*, 2. Frege, *Begriffsschrift*, x. As Uwe Dathe points out, Scholem most likely only attended the first of

Frege's lectures in Jena in April 1917; see Dathe, "Jena—Eine Episode aus Gershom Scholems Leben," 78.

45. The analogy between pure thought and infinitesimal calculus becomes evident in Cohen's sections on the "logic of origin," in *Werke*, 6:31–38 and 121–144. A discussion of a metaphorics of analogy can be found in the next chapter.

46. Other commentators translate Scholem's word *Gleichnis* here as "metaphor" (Barouch) and "image" or "semblance" (Ng); see Barouch, "Lamenting Language Itself," 6; and Ng, "'+1': Scholem and the Paradoxes of the Infinite," 200–201. I have chosen "analogy" because it not only renders legible Scholem's link to and distinction from Rosenzweig's metaphorics of analogy but also more closely matches the mathematical connotation of equation (*Gleich-nis*) with the Greek root *analogia*, meaning proportion.

47. See Aristotle, *Categories*, 86–88. In his discussion of the infinite judgment, Cohen discusses the privative *a-* in relation to the in-finite (that lacking finitude), which Rosenzweig uses to move from the nothing to the something, see Cohen, *Werke*, 6:84–89; see also Pollock, *Franz Rosenzweig*, 152–153.

48. The non-Jewish traditions such as Platonism and Gnosticism will provide a fundamental and, for critics such as Moshe Idel, problematic historical source in Scholem's account of Jewish mysticism. Idel discusses the Gnostic origins of the Kabbalah and criticism of Scholem, whose positing of the origins of Kabbalah in "non-Jewish intellectual universes" represents to Idel a more sophisticated version of *Wissenschaft des Judentums*; see Idel, *Kabbalah*, 30–32, here 30.

49. For a definition of mathematical Platonism, see Field, *Realism, Mathematics, and Modality*, 1.

50. Here Scholem takes issue with Voss's summary of Dedekind's view on the origin of numbers; see Voss, *Über das Wesen der Mathematik*, 31n3. As Dedekind writes: "My main answer to the question raised in the title of this book: numbers are the free creation of the human mind [*freie Schöpfungen des menschlichen Geistes*], they serve as a medium to grasp more easily and precisely the differentness of things" (*Was sind und was sollen die zahlen*, iii). In an entry omitted from publication, Scholem mentions receiving *Was sind und was sollen die Zahlen* in tandem with Lotze's *Logik* in October 1917; see diary entry by Scholem, October 17, 1917, in Gershom Scholem's Diaries, 59.

51. Gödel, "Some Basic Theorems on the Foundations of Mathematics and Their Implications," 323.

52. Blumenberg, *Paradigms for a Metaphorology*, 3; see also 14.

53. Scholem quotes here form the first edition of *Logic of Pure Knowledge* on page 199 (Cohen, *Werke*, 6). See also S 1:261.

54. On the potential criticism of Cohen's use of infinitesimal calculus, see my discussion in chapter 3 as well as Funkenstein, *Perceptions of Jewish History*, 289–290.

55. Scholem's discussion of the interrelations between mathematics and mysticism pervade his journals. See his comments on Novalis (S 1:265 and Scholem, *Briefe*, 1:94–95); on Buber's *Daniel* (1913) and Steiner (S 1:371); and Scholem's epistolary criticism of Goldberg (Benjamin, *Gesammelte Schriften*, 1:233–239), also discussed in Voigts, *Oskar Goldberg*, 118–122.

56. Steiner, "Mathematik und Okkultismus," 9 and 18.

57. See, for instance, S 1:360, 410, and 422.

58. Tradition has long been a critical concept in work on Scholem, see Alter, *Necessary Angels*. Other interpretations of Scholem's concept of tradition have focused on the dynamic interaction of orality and writing in the work of Franz Joseph Molitor, *Philosophie der Geschichte*, 6. See also Mertens, *Dark Images, Secret Hints*, 80–96; and Jacobson, *Metaphysics of the Profane*, 114–122.

59. See Peckhaus, "Mathematical Origins of Nineteenth-Century Algebra of Logic," 159.

60. For a general history of the development of logical calculus, mathematical logic, and modern formal logic, see Peckhaus, *Logik*, 200–214.

61. See Lotze's "Anmerkung über logischen Calcül," in *Logik*, 256–269, here 259. Cf. Boole, *Investigation of the Laws of Thought*, 49–50. For an accessible introduction to Boole's logic (and its influence on the digital computer), see Davis, *Universal Computer*, chap. 2.

62. The French mathematician Louis Couturat introduced the term *logistique* in a talk held in 1905, favoring it over "symbolic logic" and "algebra of logic"; see Grattan-Guinness, *Search for Mathematical Roots*, 367. Scholem notes (S 2:109) that the term *logical calculus* (*Logikkalkül*) is the antiquated term used by Lotze, *Logik*, 256.

63. For the intellectual conflicts among Scholem, Bauch, and Lotze, see Dathe, "Jena—Eine Episode aus Gershom Scholems Leben," 74–75. See also Bauch, "Lotzes Logik."

64. Scholem, "Bezieht sich die reine Logik . . . ," 1.

65. After holding the *Referat*, Scholem records his frustration with its in-class reception in his diary: "Held my *Referat* this morning. Bauch was indifferent, discussion worthless. Everything was silent [*alles schwieg*]" (diary entry by Scholem, November 10, 1917, in Gershom Scholem's Diaries, 32).

66. For the myriad other uses of symbols and theorization of symbolism in Scholem's intellectual career, see Idel, *Old Worlds, New Mirrors*, 83–108.

67. Benjamin, *Gesammelte Schriften*, 2.1:147–150.

68. As the character Biba explains toward the novel's semi-utopian end: "The communication of thoughts is much easier to produce through books and other written signs [*Schriftzeichen*]. If we surface-beings already know the fixed transmission of thoughts, another totally different type of understandable written signs may be commonly used among stars" (Scheerbart, *Lesabéndio*, 5:472). See also Julia Ng's work on Scholem's studies in astronomy with Wilhelm Förster, Ng, "+1: Scholem and the Paradoxes of the Infinite," 198–200.

69. Scholem to Lissauer, October 20, 1916, in Gershom Scholem Archive, 2. See also S 1:439.

70. Scholem writes: "Even in this ecstatic frame of mind, the Jewish mystic almost invariably retains a sense of the distance between the Creator and His creature. The latter is joined to the former, and the point where the two meet is of the greatest interest to the mystic, but he does not regard it as constituting anything so extravagant as identity of Creator and creature," (*Major Trends in Jewish Mysticism*, 122–123). See Idel's revisionist reading of Scholem's denial of the "extreme form of *unio mystica*" in *Studies in Ecstatic Kabbalah*, 3–4.

71. As Amir Engel points out, the idea of silence developed conceptually for Scholem as a rhetorical antidote against Buber's vocabulary of the youth movement, which emphasized chatter, experience, and action; see Engel, *Gershom Scholem*, 60. The origins of Scholem's notion of "schweigen" relate to his conversations with Benjamin and the latter's rejection of Buber's link of language and action; see Benjamin, *Gesammelte Briefe*, 1:326–327. In late 1916 and early 1917, Scholem employs "silence" to de-instrumentalize the learning of Hebrew in the Zionist youth movement as a vehicle of "national renaissance" (S 1:431). As Scholem responded in his diary to the criticism of a fellow member of the *Blau-Weiss* youth organization (Hans Oppenheim), who believed that language was one of the most important means of nationalization: "One learns Hebrew in order to be silent in Hebrew, then it is in order" (S 1:474 and here 2:15). Cf. Oppenheim, "Eine Kritik des Blau-Weiß," 12. Moreover, including *schweigen* in "the teaching" constituted Scholem's attempt to include Kraft's ambivalent response to his proclamation of interest in mathematics and Judaism during the early stages of their correspondence; see Scholem, *Briefe an Werner Kraft*, 40.

72. See the recent collection of essays on philosophical and literary approaches to lament in Jewish thought and literature in Scholem, "On Lament and Lamentation." Scholem, "List of Books in Scholem's Library," 1.

73. For an extended interpretation of the spatial metaphor of "border," see Ferber, "A Language of the Border," 176–185. Border or limit (*Grenze*) is a key mathematical concept in calculus and analytic geometry.

74. Scholem, *Major Trends in Jewish Mysticism*, 27.

75. Weigel, "Scholems Gedichte und seine Dichtungstheorie," 34–35. Scholem, "Ein mittelalterliches Klagelied," 283.

76. Budde, "Das Hebräische Klagelied," 5–6.

77. I have tried to render in English the rhythm of Scholem's translation in German as well as retain some of the syntactic oddities of his German translation.

78. Ferber astutely emphasizes the sonic nature of Scholem's lamentations citing a wealth of evidence from his diaries; see Ferber, "A Language of the Border," 176–185. Scholem writes, for instance, "lament can be contained in music, indeed in the acoustic sphere, but really without words [*wortlos*]" (S 2:139).

79. Benjamin's letter from March 30, 1918 cites "some basic lacunae and vagaries" in Scholem's theory of lament, claiming Scholem's translations "took no inspiration from the German language" (*Gesammelte Briefe*, 1:443–144). In contrast, Benjamin's own essay "The Task of the Translator" (1921) upholds Hölderlin's as a model (*Urbild*) of translation into the German. Of note is how the text warns that, even as a translator such as Hölderlin opens the doors of language, they may "slam shut and enclose the translator in silence [*ins Schweigen schließen*]" (*Gesammelte Schriften*, 4:21). On Benjamin and Scholem's theories of translation and their co-influences see Weber, *Benjamin's -abilities*, chaps. 5 and 6; Sauter, "Hebrew, Jewishness, and Love"; Eiland and Jennings, *Walter Benjamin*, 157–160; and Schwebel, "Tradition in Ruins."

80. Adorno and Horkheimer, *Briefe und Briefwechsel*, 4.1:279.

81. Ferber and Schwebel, *Lament in Jewish Thought*, sec. 5.

82. Benjamin, *Gesammelte Schriften*, 1.2:697. It is no coincidence that the epigraph to the ninth thesis "On the Concept of History," quoted here, is the sixth stanza of Scholem's poem "Greetings from Angelus" ("Gruß vom Angelus"). Scholem sent the poem to Benjamin in September 1933; see Benjamin and Scholem, *Briefwechsel 1933–1940*, 104–105.

83. Foucault, *Archaeology of Knowledge*, 6. Here I cite Foucault in order to emphasize the similarity between the theory of history implied by negative mathematics and those developed by the poststructuralists.

84. Scholem, *Major Trends in Jewish Mysticism*, 350.

85. Scholem, *On the Kabbalah and Its Symbolism*, 2.

86. Scholem, *Briefe*, 1:89.

87. On the "crisis of intuition," see Volkert, *Die Krise der Anschauung*, xxii–xxiii and 99–128.

88. Scholem was aware of these developments: "it is really a thoroughly metaphysical statement, that time would, so to say, be a line; maybe it is a

cycloid or something else, that on many points has in fact no direction (where there are no tangents)" (S 1:390). See Scholem, *Walter Benjamin*, 45. A note in Scholem's handwriting contains the Weierstraß equation, preserved in his notebook from Knopp's 1915–1916 differential equations course; see Fenves, *Messianic Reduction*, 274n20; See also Scholem, "Notebook to Knopp's Lectures on Differential Equations," vol. 1.

89. On Weierstraß's "monster" function, see Fenves, *Messianic Reduction*, 111 and 240–241.

3. INFINITESIMAL CALCULUS

1. See, for instance, the philosophical "tools" in Baggini and Fosl, *Philosopher's Toolkit*. As Leif Weatherby demonstrates, the concept of the tool, organ, and *organon* played a major role in the metaphysics of the German Romantics around 1800. On the philosophical history of the term, see Weatherby, *Transplanting the Metaphysical Organ*, 1–7.

2. Adorno, *Gesammelte Schriften*, 3:19.

3. Rosenzweig, *Gesammelte Schriften*, 2:23 (hereafter cited in text as R, followed by volume and page number). On Cohen and Rosenzweig, see Batnitzky, *Idolatry and Representation*, 17–31; Hollander, *Exemplarity and Chosenness*, 13–42; Gordon, *Rosenzweig and Heidegger*, 39–81; Gibbs, *Correlations in Rosenzweig and Levinas*, 46–54; and Fiorato and Wiedebach, "Hermann Cohen im Stern der Erlösung." On forms of representation and messianism, see Dubbels, *Figuren des Messianischen in Schriften deutsch-jüdischer Intellektueller 1900–1933*, 348–349.

4. Interpretations of Rosenzweig's use of mathematics have often focused only on its role in *The Star of Redemption*. Most often, mathematics provides the means of generating the "something" from the "nothing"; see Pollock, *Franz Rosenzweig*, 150–152; and Gibbs, *Correlations in Rosenzweig and Levinas*, 36 and 48. Dana Hollander adds that the infinitesimal provides Rosenzweig with a way of conceiving of Jewish election; see Hollander, *Exemplarity and Chosenness*, 27–39. Perhaps most revealing about the relatively marginal attention that mathematics has received in Rosenzweig scholarship is the fact that his use of mathematics has been described dismissively as an "allegory," "analogy," or a "metaphor"; see, respectively Gordon, *Rosenzweig and Heidegger*, 39–51; Gordon, "Science, Finitude, and Infinity," 44–46; and Funkenstein, *Perceptions of Jewish History*, 288–290.

5. See Rosenzweig's paradigmatic formulation: after the war, "a field of ruins marks the place where the [German] empire once stood" (*Hegel und der Staat*, 17–18). See also Susman, "Der Exodus aus der Philosophie." Adorno cites the dissolution of the totality of philosophical systems in his *Antrittsvorlesung* as

forcing the question of the "actuality of philosophy" (*Gesammelte Schriften*, 1:325–327).

6. See Pollock, *Franz Rosenzweig*, especially 1–14.

7. See R 1.1:132–138, as well as Pollock, *Franz Rosenzweig's Conversions*, chap. 2. See also Glatzer, *Franz Rosenzweig*. Benjamin Pollock offers more compelling accounts of both the origins and implications of Rosenzweig's *Star of Redemption* and his near-conversion following the all-night conversation in 1913 among Rosenzweig, Eugen Rosenstock, and Rudolf Ehrenberg. See Pollock, *Franz Rosenzweig's Conversions*; and *Franz Rosenzweig*.

8. Benjamin, *Gesammelte Schriften*, 2.1:75. Adorno's *Minima Moralia* ends with a call to "view all things as they appear from the standpoint of redemption" (*Gesammelte Schriften*, 4:283). On messianism in Adorno and Benjamin, see Brunkhorst, *Adorno and Critical Theory*, 44–45.

9. Derrida, *Specters of Marx*, 74. For a discussion of the relationship between Rosenzweig and Derrida's messianisms, see Kavka, *Jewish Messianism*, 129–158, 196–197. See also Schulte, "Messianism Without Messiah."

10. Rabinbach, *In the Shadow of Catastrophe*, 33. Hans Blumenberg's claim that Jewish mysticism projects the apocalypse "beyond history," in *The Legitimacy of the Modern Age*, 41. On the forms of modern Jewish messianism, see Dubbels, *Figuren des Messianischen in Schriften deutsch-jüdischer Intellektueller 1900–1933*, on Rosenzweig, 41–72 and 347–348. As discussed in chapter 2, this deferral of reconciliation makes up for Scholem the paradox of mysticism; Scholem, "Messianic Idea in Judaism," 35.

11. Adorno, *Gesammelte Schriften*, 4:283.

12. This account of differentiation and integration is taken from Riecke, *Lehrbuch der Experimental-Physik*, 17–19. Rosenzweig took Eduard Riecke's physics course in the summer semester of 1905, see Rosenzweig, "Abgangs-Zeugnis"; and *Verzeichnis der Vorlesungen*, 8. On the differences between Leibniz and Newton, see Guicciardini's "Newton's Method and Leibniz's Calculus" in Jahnke, *History of Analysis*, 73–104. On Newton and Leibniz, see also Boyer, *History of the Calculus*, 187–223.

13. Figures 3.1 and 3.2 reprinted from Riecke, *Lehrbuch der Experimental-Physik*, 17–18.

14. On German Idealism, see the work of Frederick Beiser, for example, *Fate of Reason*. On Rosenzweig's studies of German Idealism, see Pollock, *Franz Rosenzweig*.

15. On the history of reforms in mathematical pedagogy in Germany, see Pyenson, *Neohumanism*, chap. 6; Schubring, "Pure and Applied Mathematics in Divergent Institutional Settings in Germany." It may seem odd to speak of "modernity" in terms of infinitesimal calculus around 1900; indeed, historians

of mathematics such as Herbert Mehrtens have more readily associated "modernity" in mathematics with an emphasis on symbolic rigor and logical consistency as advocated by Karl Weierstraß and David Hilbert and a "counter-modernity" with the characteristics of intuitiveness and utility. See Mehrtens, *Moderne*, chaps. 2 and 3.

16. On the cultural history of geometry as a model of logical order, see Alexander, *Infinitesimal*. See also Spinoza's *Ethics* (1677), whose subtitle reads "ordine geometrico demonstrate [demonstrated in geometrical order]."

17. Pyenson, *Neohumanism*, 13. Schubring adds that "in line with the neohumanist values at these universities [in Prussia], particularly in the philosophical faculties, mathematics became established as a 'pure' science, as 'pure mathematics.' Because teaching at *Gymansien* was initiated as a scientific profession with a high standard of training in order to ensure the social status of teachers, and because mathematics teachers also represented and practiced the new scientism, their activity encouraged the spread of the ethos of pure mathematics" (*Conflicts*, 484).

18. Klein, *Vorträge über den mathematischen Unterricht an den Höheren Schulen*, 6. See also the Meraner reforms, in August Gutzmer's *Reformvorschläge für den mathematischen und naturwissenschaftlichen Unterricht*, 5. See also Schubring, "Pure and Applied Mathematics in Divergent Institutional Settings in Germany," especially 175–180; and "Mathematics Education in Germany (Modern Times)," 247–249.

19. Here I summarize the debate between Klein and the German-Jewish mathematician (and Thomas Mann's father-in-law) Alfred Pringsheim that arose in 1898 over the method of introducing students to infinitesimal calculus. Quoted in Bergmann, "Mathematics in Culture," 194–195.

20. Rosenzweig will return to the term "organ"/"organon" as a metaphor for mathematics in his discussion of Hermann Cohen in *The Star of Redemption*, see Cohen's references to the infinitesimal as "organon" in *Werke*, 5:4 and 133. Cohen also uses the terms "instrument" (ibid., 6:32) and "tool" (7) to refer to the infinitely small. The Romantic poet, Novalis, also refers to mathematics as an "organ," meaning the objectification of the understanding in the context of the greater project that Leif Weatherby calls Romantic organology; see Weatherby, *Transplanting the Metaphysical Organ*, 215.

21. Abstraction, purity, and detachment were the metaphors in which geometry had been valued in neohumanistic education, see Pyenson, *Neohumanism*, 5 and 13. They are also the terms in which geometry came to be devalued later in the nineteenth century; in the words of Baumeister, "long before Darwin's theory ruled the descriptive natural sciences, modern geometry had broken with the rigidity [*Starrheit*], that is, the immutability of Euclid's

figures and found . . . the royal road [*den Königsweg*] to geometry, of which Euclid did not know" (*Handbuch der Erziehungs- und Unterrichtslehre für höhere Schulen*, 72). The essay "*Volksschule* and *Reichsschule*," cites as its source Nernst and Schönflies, *Einführung in die mathematische Behandlung der Naturwissenschaften*; see R 3:392 and Waszek, *Rosenzweigs Bibliothek*, 113.

22. Schopenhauer, *World as Will and Representation*, 1:95.

23. Rosenzweig misidentifies the dialogue in "*Volksschule* and *Reichsschule*" as Plato's *Theaetetus* (R 3:389). Cf. Plato, *Meno and Other Dialogues*, 114–123. Chamberlain also cites the example of *Meno* in his lecture on Plato; the slave demonstrates how "one can understand geometry, without having learned it" (*Immanuel Kant*, 515). Likewise, Mendelssohn mentions the *Meno* example in "On the Evidence in Metaphysical Sciences," 258 (see chapter 1).

24. Susman, "Der Exodus aus der Philosophie." Ironically, it is precisely the round-about idealistic and systematic nature of Rosenzweig's work that Kracauer remarks on when discussing *The Star of Redemption* in a letter to Löwenthal: "Rosenzweig's book is significant, clearly systematic drivel that kills Idealism for us, only to reestablish it after the fact" (*In steter Freundschaft*, 44).

25. In his *Grundlage der gesamten Wissenschaftslehre* (1794/1795), Fichte takes the principle of identity ("A is A" or "A = A") to be the "perfectly certain and established" foundation of thought. Here Fichte argues that this proposition would be impossible if there were not a subject ("I am") to think it; see Fichte, *Science of Knowledge*, 94–102. "A= A" also symbolized in the philosophy of mathematics the idea that all mathematical knowledge is in essence analytic—that is, can be derived from the simple tautology, A=A (see chapter 2).

26. See, for instance, §175 in Goethe's *Zur Farbenlehre*, 80–81.

27. Rosenzweig, *Die "Gritli"-Briefe*, 124.

28. See Smith, "Infinitesimal as Theological Principle," 563.

29. Rosenzweig's letters to his parents make clear that Rosenzweig has in mind his experience in geometry while a student at the *Gymnasium* with "Herr Prof. Hebel"; see Rosenzweig, *Feldpostbriefe*, 304 and 403.

30. See, for instance, Horkheimer, *Eclipse of Reason*, 3.

31. Scholars have tended to cast the stakes of Rosenzweig's 1913 near conversion in terms of a battle between "faith" and "science"/"reason," which, it would seem, would have particular bearing on this analysis; see, for instance, Glatzer, *Franz Rosenzweig*, xi. As Pollock suggests, at stake for Rosenzweig in the confrontation with Rosenstock was less "faith" and much more "the moral or spiritual status of the world" and the relationship between self and a redeemable or irredeemable world (*Franz Rosenzweig's Conversions*, 3).

32. On the diverse and interdisciplinary influences on Rosenzweig's thought, see the contributions in Brasser, *Rosenzweig als Leser, Kontextuelle Kommentare zum "Stern der Erlösung."*

33. As Rosenzweig explains at the outset of the paragraph in which the analogy is contained: "At this point I realize that, as I want to continue writing, everything that I now would have to write to you is, for me, inexpressible [*unaussprechbar*] to you," (R 1.1:283).

34. I cite here from Newton's text, translated by Gerhard Kowalewski, that Rosenzweig requested his mother send to him while in Macedonia in October 1916; see *Feldpostbriefe*, 273n288. See also Newton, *Newtons Abhandlung*, 4. Further evidence that Rosenzweig had this book in mind is that, despite the terminological difference between Leibniz and Newton, Kowalewski's commentary employs the term "differential quotient" to explicate the ratio between fluxions; see ibid., 4n3.

35. Ibid., 3. On Leibniz's belief in the existence of infinitesimals, see Leibniz, "Letter to Varignon," 543–546; and the contributions in Goldenbaum and Jesseph, *Infinitesimal Differences.*

36. Figure 3.3 reprinted from Newton, *Newtons Abhandlung*, 4.

37. Newton writes: "Determining the fluents from the fluxions is a difficult problem; the initial step of the solution is equivalent to the quadrature of the curve" (ibid., 7).

38. Bradshaw, Marcus, and Roach, *Moving Modernisms*; Rabinbach, *Human Motor*, chaps. 2 and 4.

39. See Bergson, *Creative Evolution*, 87–97.

40. As Rosenzweig writes to his parents (R 1.1:271) and Rudolf Ehrenberg (321).

41. Rosenzweig admits, already in 1905, that he will have to put on "blinders" to Chamberlain's idea of race, if he hopes to cultivate himself (R 1.1;18 and 40). Even the idea of movement, life, and dynamism, which Rosenzweig draws on from Chamberlain, is the element of European culture for the latter, which must be protected against "bestial barbarism" of the East (Russian and Asia) and South (Africa). See Chamberlain, *Immanuel Kant*, 701. For more on Chamberlain, see Field, *Evangelist of Race.*

42. Chamberlain, *Immanuel Kant*, 702. As Anja Lobenstein-Reichmann explains, these terms circulated much more broadly as markers for the rising bourgeoisie; see Lobenstein-Reichmann, *Houston Stewart Chamberlain*, 247–249, here 249.

43. Chamberlain, *Immanuel Kant*, 703.

44. Here Rosenzweig refers to the collection of Aquinas's texts published by Engelbert Krebs in 1912 as Aquinas, *Texte zum Gottesbeweis*; see R 1.1:273.

45. Interestingly, the material form of this passage embodies the link among belief, subjectivity, and motion, belonging to a set of theological and philosophical notes, a "Paralipomena," that Rosenzweig wrote on postcards while stationed in Macedonia and sent home to Kassel. As Rosenzweig wrote to his mother: "Now and then I will write letters addressed to 'Dr. Rosenzweig, c/o Counselor of Commerce Rosenzweig Terasse 1.' These contain *only* scientific notes and you are *not* allowed to open them. Rather, keep them safe, best with extra numbering, so that I have them all together later" (R 1.1:184).

46. As elucidated in book 3 of Aristotle, *Physics*, 57. Thomas Aquinas, who Rosenzweig credited with his rediscovery of infinitesimal calculus, defines motion as "to move another is nothing other than to bring something from potentiality to actuality." Since things are in motion and cannot set themselves in motion, there must be an actuality that originally set them into motion, which Thomas takes as proof of God. See *Summa Theologicae* in Aquinas, *Basic Works*, 53. This section is reproduced in Krebs's selection Aquinas, *Texte zum Gottesbeweis*, 53–54.

47. On the different types of motion in Aristotle and their origins, see *On the Heavens*, esp. bk. 1, chap. 2. See also Aristotle, *Physics*, chap. 8, esp. 229.

48. Newton, "Principia," 70.

49. Here, Aristotle's arguments responded to the Atomists, see book 6 of Aristotle, *Physics*, 138–152; see also Aquinas, *Commentary on Aristoteles's Physics*, 388–390.

50. Euclid, *Thirteen Books of Euclid's Elements*, 1:153.

51. Aristotle, *Physics*, 161–162. Zeno's paradoxes are fourfold, including the paradox of Achilles and the tortoise.

52. Kant, *Critique of Pure Reason*, A169–170/B211. Translation modified to fit Rosenzweig's terminology. On Kant's reference to Newton, see Friedman, *Kant and the Exact Sciences*, 74–75.

53. I cite here only in part from the schema of numbers, which also include "infinitesimal numbers," that Rosenzweig's 1918 letter proposes. See Mosès, *Angel of History*, 52–55.

54. December 16, 1921, cited in Belke and Renz, *Marbacher Magazin*, 47:36.

55. As Shlomo Pines argues, Rosenzweig's distasteful dismissal of Islam was based primarily on his reading of Hegel and Hegel's account of progress in Islam; see R 2:251–254 and Pines, "Der Islam im 'Stern der Erlösung.'" On the tenuousness of Rosenzweig's argument regarding Islam as well as the histories of Greece, China, and India, see Gibbs, *Correlations in Rosenzweig and Levinas*, 113–117.

56. As Adorno writes regarding *The Star of Redemption*: "I wouldn't understand it even if I were to understand it. Nonetheless, I want to read the book, because it gets at the most important things"; quoted in Kracauer and Löwenthal, *In steter Freundschaft*, 44–45.

57. Buck-Morss, *Origin of Negative Dialectics*, 5–6.

58. On Rosenzweig and Cohen, see Batnitzky, *Idolatry and Representation*, 17–18 and 29.

59. Ibid., 29.

60. Rosenzweig is explicit about the link between Chamberlain and Cohen, claiming that his use of mathematics is "indirectly Marburg'ian" in reference to the "Marburg School" founded by Cohen. (R 1.1:321). Likewise, he even describes Chamberlain as an "orthodox Cohen'ian," who "burns what Cohen burns (i.e., Aristotle, Thomas, Hegel, Schopenhauer) and worships what Cohen worships (i.e., Plato, Leibniz, Schiller)" (456).

61. Cohen, *Werke*, 6:13. See the opening of Kant, *Critique of Pure Reason*, B1. On the origins of Neo-Kantianism, see Kohnke, *Rise of Neo-Kantianism*.

62. Cohen, *Werke*, 6:28. On Cohen's "logic of origin," see ibid., 6:31–38; on the "judgment of origin," see ibid. 84–89. For Kant, the infinite differs from the negative judgment, which takes the form "x is not A," and as Paul Guyer explains, "merely denies a predicate of the concept of the object." The infinite judgment takes the form "x is non-A" and "*asserts* 'non-A' of its subject." It "leaves open an infinite range of predicates for *x*, but implies that *some* predicate applies to it" (*Kant*, 76).

63. Cohen, *Werke*, 6:33.

64. Ibid., 6:125.

65. See ibid., 6:135 and 35.

66. See Albrecht, "'[H]eute gerade nicht mehr aktuell,'" 21–24.

67. Pollock, *Franz Rosenzweig*, 1 and 12.

68. See Hegel, *Science of Logic*, 59. Cohen employs the concept of the "relative" nothing in contrast to general "non-being," see Cohen, *Werke*, 6:93–94. See also Gibbs, *Correlations in Rosenzweig and Levinas*, 49–51.

69. Terminologically, Rosenzweig draws here on Abel, *Über den Gegensinn der Urworte*. See Rosenzweig, *Die "Gritli"-Briefe*, 253. Freud draws on Abel as well in *Totem and Taboo* (1913); Jacques Derrida also makes use of Abel in his discussion of "undecidability" in *Dissemination*, 220.

70. For Rosenzweig's derivation of "God's freedom," see R 2:32: "God's freedom is a violent no par excellence." On "God's vitality," see ibid., 34.

71. See Gibb's depiction of Rosenzweig's grammar in *Correlations in Rosenzweig and Levinas*, 67.

72. Santner, *On the Psychotheology of Everyday Life*, 5.

73. For Cohen, after Newton, "the point means something different, something more positive" (*Werke*, 6:129).

74. Scholem, "On the 1930 Edition of Rosenzweig's *Star of Redemption*," 323.

75. See Benjamin, *Gesammelte Schriften*, 2.1:694. For Horkheimer, Cohen's equation of an "eternal logos" with "the snippet of the world presented to the scholar, which becomes ever more expressible in the form of differential quotients" is the "false self-consciousness of the bourgeois scholar in the liberal era" (*Gesammelte Schriften*, 4:171–172).

76. The concept of *Bewährung*, usually translated as "confirmation" or "verification," also plays a role in Max Weber's *The Protestant Ethic and the Spirit of Capitalism* (1905) and in Weber's essay "Die protestantischen Sekten und der Geist des Kapitalismus," 234–235. In Rosenzweig, verification has often been interpreted along the lines of martyrdom; as Martin Kavka writes: "Martyrdom, and perhaps only martyrdom, is what verifies the truth of one's belief" ("Verification (Bewährung) in Franz Rosenzweig," 176); see also Batnitzky, *Idolatry and Representation*, 44.

77. At least semantically, Rosenzweig's idea of "verification" anticipates Carnap's "Truth and Verification" ("Wahrheit und Bewährung," 1936). For both Rosenzweig and Carnap, knowledge demands empirical "verification." Not only Carnap but also Hans Reichenbach and Karl Popper employ the term *verification* to signify how a theory can be empirically "verified" (or "corroborated" in Popper's case) as true. See, for instance, Popper, *Logik der Forschung*, vol. 3, chap. 10. Adorno's description of Reichenbach's theory of verification (*Bewährung*) from the Paris Congress of the Unity of Science, mentioned in chapter 1, comes uncannily close to Rosenzweig's; see Adorno and Horkheimer, *Briefe und Briefwechsel*, 4.1:570.

78. Rosenzweig to Kracauer, May 25, 1923, published in Baumann, "Drei Briefe," 174–175. A few sentences later, Rosenzweig explains: "Now, critical for your 'state of mind' is if you realize how what I just explained to you is *not* a form of subjectivism, but rather, on the contrary, it is the path upon which the ideal of objectivity can really be universally realized—an ideal that was always proposed by the old [logic] (old being Aristotle *as well as* Kant), but really only strictly enforced for one narrow region, basically only for 'mathematics' [*das 'Mathematicshe'*]," 175. On the relationship between Rosenzweig and Kracauer, see also Handelman, "The Forgotten Conversation."

79. See, for instance, Frege, *Foundations of Arithmetic*, 1: "After deserting for a time the old Euclidean standards of rigour, mathematics is now returning to them, and even making efforts to go beyond them. . . . Proof [*Beweis*] is now demanded of many things that formerly passed as self-evident."

80. Rosenzweig writes to Rosenstock in 1925: "You thus see: in the Lehrhaus, I've already let the Star shine" (R 1.2:1027). Rosenzweig's work at the Lehrhaus was not only central to postwar constructions of German-Jewish identity but also, for Glatzer, an extension of Rosenzweig's work begun already in *"Volksschule* and *Reichsschule"*; see Glatzer, "Frankfort Lehrhaus," 155.

81. Dimock, *Through Other Continents,* 75–76 and 87.

4. GEOMETRY

1. Siegfried Kracauer to Margarete Susman, January 11, 1920. On Kracauer's relationship with Susman, who first helped him establish connections with the *Frankfurter Zeitung*, see Belke, "Siegfried Kracauer."

2. Adorno, *Gesammelte Schriften,* 1:332.

3. See, for instance, Adorno's depiction in "The Curious Realist" ("Der wunderliche Realist," 1964), here *Gesammelte Schriften,* 11:387. Histories of critical theory often take note of Kracauer's early influence on Adorno; see, for example, Brunkhorst, *Adorno and Critical Theory,* 21–23. Indeed, Kracauer's relationship with Adorno redefines the limits of intellectual friendships; see von Moltke, "Teddie and Friedel." More recently, Kracauer has been increasingly recognized for his significant contributions to the development of film theory as well as his position in the postwar intellectual milieu in the United States; see von Moltke, *Curious Humanist;* and Hansen, *Cinema and Experience,* especially pt. 4. Regarding Kracauer's reception, see the contributions from Miriam Hansen, Gertrud Koch, and Heide Schlüpmann to the special issue of *New German Critique,* no. 54 (2011); and Gemünden and Moltke, "Introduction: Kracauer's Legacies." On Kracauer's biography, see Belke and Renz, *Marbacher Magazin;* Gilloch, *Siegfried Kracauer;* and Später, *Siegfried Kracauer.* On Kracauer and the intellectual experience of the Weimar Republic, see Craver, *Reluctant Skeptic;* and Schröter, "Weltzerfall und Rekonstruktion."

4. Here I follow a point made by Johannes von Moltke and Gerd Gemünden: "[Kracauer's] cultural critique [was] distinct, on the one hand, from cultural criticism, which would exhaust itself in the attentive treatment of individual objects; and on the other hand, from cultural theory, which would join philosophy in its construction of overarching concepts for defining the very notion of culture itself" ("Introduction: Kracauer's Legacies," 3).

5. See, for instance, von Moltke, *Curious Humanist,* 124; and Huyssen, *Miniature Metropolis,* 130–135.

6. Hansen, *Cinema and Experience,* 6. Kracauer makes his early enthusiasm for Lukács's historical narrative clear in his review of *Theory of the Novel*

(1921); see Kracauer, *Werke*, 5.1:282–288. See also *Sociology as Science* (*Werke*, 1:12). The idea of the disjointedness of thought from the material world runs throughout Kracauer's career. For example, in *History: The Last Things Before the Last* (1969): "Human affairs" he writes, "transcend the dimension of natural forces and causally determined patterns," assuring that any approach to history or sociology "which claims to be scientific in a stricter sense of the word will sooner or later come across unsurmountable [*sic*] obstacles" (29).

7. This humanist politics and the hope to create a society based on human needs and through the participation of human actors runs from Kracauer's Weimar writings to his later work on film and history; for a detailed discussion of these politics, see von Moltke, *Curious Humanist*.

8. Kracauer, *Werke*, 5.2:615 (hereafter cited in text as K, followed by volume and page number).

9. The earliest work on Horkheimer and Adorno's critical theory—for instance, by Jay and Buck-Morss—points out the significance of aesthetic mediation in, especially, Adorno's thought. See Jay, *Dialectical Imagination*, chap. 6; and Buck-Morss, *Origin of Negative Dialectics*, chap. 8. For current perspectives on the state of Adorno's aesthetic project, see Hohendahl, *Fleeting Promise of Art*.

10. Kant invokes a similar distinction in the name "the transcendental aesthetic," whose designation of space and time as pure forms of intuition Kracauer invokes in *Sociology as Science*; see Kant, *Critique of Pure Reason*, A21/ B35–36.

11. Bubner offers a critical evaluation of Adorno's aesthetics of theory thesis in the essay "Kann Theorie ästhetisch werden?"

12. This sense of an observer and recorder of modernity marks much of Kracauer's early work. On Kracauer's early engagements with the cinema, see, for instance, Hansen, *Cinema and Experience*, 3–39. For an examination of the interplay of Kracauer's theory of modernity with philosophy and literature, see Mülder-Bach, *Siegfried Kracauer*, esp. 19–55. The feeling of being an "outsider" both to the Frankfurt School and the modern society that he sought to analyze permeates both Kracauer's work as well as its reception, perhaps most legible in the notion of "extra-territoriality"; see Kracauer, *History*; Koch, "Not Yet Accepted Anywhere," 95; and Jay, "Extraterritorial Life of Siegfried Kracauer," 153.

13. Husserl, *Ideas*, 140.

14. When Kracauer started his studies at the Technische Hochschule in Darmstadt in the summer semester of 1907, he took "Darstellende Geometrie" with Reinhold Müller (Belke and Renz, *Marbacher Magazin*, 47:7). After transferring to Berlin, Kracauer took "Darstellende Geometrie" with Erich

Salkowski in the winter semester 1907/1908. In October 1907, Kracauer notes in his journal: "In the afternoon, I had descriptive geometry. How insufferable these things are to me!" quoted in ibid., 11.

15. See Descartes, *Discourse on Method*, 104–106.

16. On the history and impact of Taylorism in Germany, see Nolan, *Visions of Modernity*, esp. 123.

17. Kracauer is explicit about his use of Husserl's *Ideen zu einer reinen Phänomenologie*. The title of *Sociology as Science* refers to Husserl's 1911 essay "Philosophy as Strict Science" ("Philosophie als strenge Wissenschaft").

18. Euclid, *Thirteen Books of Euclid's Elements*, 1:221–240. Aristotle provides a sharper definition of axioms: in contrast to a postulate, which is not necessarily true in itself, an axiom is convincing in itself and serves as, "that which must be grasped if any knowledge is to be acquired" (*Posterior Analytics*, 33 and 7). On the varying use of the terms *axioms* and *postulate* in Aristotle and Euclid, see Merzbach and Boyer, *History of Mathematics*, 94–96.

19. On the connection between Simmel and Kracauer, see Frisby, *Fragments of Modernity*; and Goodstein, *Georg Simmel and the Disciplinary Imaginary*, 153.

20. Simmel, *Philosophy of Money*, 114. Kracauer's critique of Simmel's thought in "Georg Simmel" (1919) upholds the "unconditional validity" of the "laws of logic, the mathematical axioms," summarizing Simmel's position on the geometric axioms: "differently organized beings would have surely come to different dogmas" (K 9.2:237).

21. Husserl, *Ideas*, 18–19.

22. See Hilbert, "Axiomatisches Denken."

23. For Kracauer, see K 1:47, 48, 46, and 39. Husserl uses the term "mathematical manifold" (*Ideas*, 140). Kracauer may have taken the image of the "truncated cone" from Husserl, who mentions it in his discussion of "mathematical manifold." This terminology reappears in Husserl, *Crisis of European Sciences and Transcendental Phenomenology*, 45.

24. Kracauer, *History*, 16.

25. Husserl, *Ideas*, 140.

26. See Euclid, *Thirteen Books of Euclid's Elements*, 1:221, 223, and 314–315. "Common notion 2" posits that equals added to equals are equals; proposition 29 states that "a straight line falling on parallel straight lines makes the alternate angles equal to one another, the exterior angle equal to the interior and opposite angle, and the interior angles on the same side equal to two right angles." In turn, the proof for proposition 29 draws exclusively on propositions 13 and 15, common notions 1 and 2, and postulate 5 (ibid., 1:311–312).

27. Husserl uses "'geometry' of experience" in *Ideas*, 138.

28. Mülder-Bach, *Siegfried Kracauer*, 35.

29. Lukács makes a similar argument in *History and Class Consciousness* (see 110–121), which was published a year after *Sociology as Science* (see chapter 1). Hansen also notes the performative dimension in Benjamin's writing; see Hansen, *Cinema and Experience*, 89.

30. Adorno, *Gesammelte Schriften*, 7:393.

31. As Adorno advises Horkheimer in a letter leading up to "The Latest Attacks on Metaphysics": "The principle impossibility of unifying [logical positivism's] twin foundational operations, experiment and calculus, is the primordial antinomy of mathematical logic [*Logistik*]" (*Briefe und Briefwechsel*, 4.1:239–240).

32. Kracauer and Löwenthal, *In steter Freundschaft*, 49.

33. Beyond his early friendships with Adorno and Benjamin, Kracauer's interest in film and his aphoristic style also influenced the members of the Frankfurt School. See Gemünden and Moltke, "Introduction: Kracauer's Legacies," 6; and Koch, *Siegfried Kracauer*, 37–38.

34. Kracauer to Susman, January 11, 1920.

35. Projection serves as a salient philosophical metaphor for thinkers such as Freud, indicating the process by which beliefs and anxieties become manifest in a form other than their own. Where an individual suppresses, distorts, and becomes conscious of an "internal perception," as Freud writes in the case of Daniel Paul Schreber, "in the form of an external perception" (*Standard Edition*, 12:65–66).

36. Boyer and Merzbach, *History of Mathematics*, 483–487.

37. Salkowski, *Grundzüge der darstellenden Geometrie*, 1. As much as projection is a practical and heuristic technique meant for professionals not specializing in mathematics, the presence of this form of geometry in Kracauer's intellectual project has the reform movements in mathematical pedagogy to thank as does the presence of infinitesimal calculus in Rosenzweig's work.

38. Figure 4.1 reprinted from ibid., 6.

39. See the descriptions of Ginster's work in the office of Herr Valentin and especially his design of a military cemetery (K 7:62–69 and 111–115). See also Gilloch, *Siegfried Kracauer*, 6 and 60.

40. Kracauer and Löwenthal, *In steter Freundschaft*, 49.

41. Kracauer's reference to the term *projection* has often been interpreted in terms of Simmel; as Mülder-Bach writes "Kracauer makes recourse" through the term projection "to the concept that Georg Simmel introduced into sociology"; see Mülder-Bach, "Nachbemerkung" (K 1:380n20).

42. Kracauer's term "Projektionslehre" matches the technical term that serves as the title of textbooks such as Schudeiskÿ, *Projektionslehre*, 1–2.

43. Kracauer to Susman, January 11, 1920. As David Frisby notes, the original title of the first chapter in the *The Detective Novel* was "The Transformation of Spheres" (*Sphärentransformation*), not just "Spheres"; see Frisby, "Zwischen den Sphären," 40.

44. See Mülder-Bach, *Siegfried Kracauer*, 38–45.

45. Kierkegaard, *Concluding Unscientific Postscript*, 246–251.

46. Parallel projections project an object from an infinitely distant point, preserving distance in projection; cavalier, in contrast, distorts distance in the projected image. See Salkowski, *Grundzüge der darstellenden Geometrie*, sec. 1.

47. On the social ramifications of the Mercator projection, see Monmonier, *Rhumb Lines and Map Wars*, chap. 2.

48. Figure 4.2 reprinted from Salkowski, *Grundzüge der darstellenden Geometrie*, 2.

49. There is always a loss of a dimension with projection; see Salkowski, *Grundzüge der darstellenden Geometrie*, 1.

50. Vidler, *Warped Space*, 72.

51. See the descriptions in K 1:130, 133, 135, and 138.

52. As in *Sociology as Science*, *The Detective Novel* draws on Kant's use of the term *aesthetic* as the point of interplay between experience and cognition mediated by space and time. Kant, *Critique of Pure Reason*, A21/B35–36.

53. Kracauer mentions these ideas to Löwenthal in letters from October 1923 and March 1922, respectively; see Kracauer and Löwenthal, *In steter Freundschaft*, 38 and 49. He references a "metaphysics of film" (*die noch ungeschriebene Metaphysik des Films*) in a review of the film *Die närrische Wette des Lord Aldini* (1923; K 6.1:43).

54. Kracauer, *History*, 4.

55. Mülder-Bach first names Kracauer's methodology "surface-level analysis" (*Oberflächenanalyse*) in reference to the opening passage from "The Mass Ornament," although Kracauer emphasizes the theoretical significance of the "surface" as early as his essay on Simmel (K 9.2:171); see Mülder-Bach, *Siegfried Kracauer*, 88. Other interpreters have emphasized the conceptual affinity of the "surface" with psychoanalysis and Kracauer's 1926 review ("Die Denkfläche") of Paul Oppenheim's book *Die natürliche Ordnung der Wissenschaften* (1926); see Hansen, *Cinema and Experience*, 51; and Koch, *Siegfried Kracauer*, 28–31.

56. The shift from the fall-and-decline narrative of modernity to a dynamic narrative of history as change and as changeable, which accompanies his readings of Marx in the early 1920s, characterizes the reception of his early work, see Mülder-Bach, *Siegfried Kracauer*, 64.

57. See chapter 1 for an analysis of Lukács's work on mathematics and its influence on critical theory.

58. Lukács, *History and Class Consciousness*, 83.

59. Ibid., 88 and 105; translation modified.

60. Ibid., 92.

61. For a detailed discussion of the ideological and artistic implications of the Tiller Girls, see Hewitt, *Social Choreography*, 177–213. Kracauer reviewed a number of such dance revues, not just the Tiller Girls; see "Die Revuen" (K 5.2:313–317), "Revue Confetti" (366–368), "Revue Nr. 1 der Wintersaison" (488–489), and "Das Berliner Metropoltheater im Schumanntheater" (542–544).

62. Gertrud Koch astutely emphasizes theological undercurrents of the term *mass*, which stems from the Hebrew *mazza* ("matzoh," unleavened bread); see *Siegfried Kracauer*, 26.

63. Scholarship on Kracauer has not been wont to point out the similarities between "The Mass Ornament" and *Dialectic of Enlightenment*; see Mülder-Bach, *Siegfried Kracauer*, 66; and Levin, "Introduction," 19. Von Moltke shows how the methodology of Kracauer's first book on film, *From Caligari to Hitler*, also resurfaces in *Dialectic of Enlightenment*; see von Moltke, *Curious Humanist*, 256n57.

64. Lukács, *History and Class Consciousness*, 139.

65. In 1925, Kracauer published the essay "Die Bibel auf Deutsch" which criticized a new translation project of the Hebrew Bible, *Die Schrift*, started by Rosenzweig and Buber in the 1920s and completed by Buber in the 1960s. See Jay, "Politics of Translation."

66. von Moltke, *Curious Humanist*, 137. On Kracauer's "double-edged" attitude toward distraction, see North, *Problem of Distraction*, 158–162.

67. Adorno, *Gesammelte Schriften*, 3:13.

68. Kracauer to Unseld, December 3, 1962, reprinted in Kracauer and Adorno, *Briefwechsel*, 571.

69. See Cesàro, *Vorlesungen über natürliche Geometrie*, esp. 1–21; Boyer, *History of Analytic Geometry*, 229–230. I have been unable to find archival evidence linking Kracauer and Cesàro; moreover, Cesàro's term "intrinseca" differs from the sense of "natural" (*naturale* in Italian) used by Kracauer and Descartes.

70. Heinz Brüggemann makes the connection between Kracauer's term and Descartes's *Optics*; see Brüggemann, *Das andere Fenster*, 282. For a detailed explication of Descartes's theory of perception, and the position of "géométrie naturelle," see Kutschmann, *Der Naturwissenschaftler und sein Körper*, 224–255.

71. Descartes, *Discourse on Method*, 106.

72. Ibid., 111.

73. Ibid., 106.

74. See ibid., 66–67.

75. Hansen, "With Skin and Hair," 447.

76. Descartes, *Discourse on Method*, 104.

77. Brüggemann writes: "Descartes gets rid of the old idea of the interlinkage of the human with the world; but he also makes possible the first steps toward a new, intellectually sovereign relationship between world and human," (*Das andere Fenster*, 282–283); Kutschmann, *Der Naturwissenschaftler und sein Körper*, 250–251. On Kant's "Copernican revolution," see Guyer, *Kant*, 51–140.

78. The Marseille texts stem from trips Kracauer took to the city in 1926 with Lili Kracauer to visit Walter Benjamin; see Kracauer and Benjamin, *Briefe an Siegfried Kracauer*, 33 and 44. See also Brüggemann, *Das andere Fenster*, 280 and 298n10. What Kracauer refers to as the "Place de l'Observance" could refer to what is today La Vieille Charité, which is flanked by Rue de l'Observance and was named Place de l'Observance before a hospital was constructed there in the seventeenth century; *Marseille à la fin de l'Ancien Régime*, 82.

79. Huyssen, *Miniature Metropolis*, 18 and 135.

80. As Benjamin describes it, Poe's short story is an "x-ray of a detective novel"; in it, an unnamed narrator follows an "unknown man who manages to walk through London in such a way that he always remains in the middle of the crowd." Benjamin, *Gesammelte Schriften*, 1.2:550. See also Poe, *Selected Tales*, 84–91.

81. raca, "Zwei Flächen," 2–3. "Two Planes" begins in the feuilleton section under the article "Wege zur Rationalisierung" by Otto Kienzle, part twelve of a series of articles dedicated to economic rationalization.

82. According to von Moltke, Kracauer's "curious humanism" was a nonanthropocentric vision of the world arranged around the subjectivity of the spectator; see von Moltke, *Curious Humanist*, 181–182.

83. Huyssen bases his reading of "The Quadrangle" on its noteworthy opening sentence: "Whoever the place finds did not seek it"; see Huyssen, *Miniature Metropolis*, 132–133. See also K 1:206–209.

84. Levin, *Siegfried Kracauer*, 165–185. See also Stalder, *Siegfried Kracauer*, 79–81.

85. See Benjamin, *Gesammelte Schriften*, 1.1:610.

86. Kraus, *Schriften*, 4:185 and 189.

87. Scholars have often noted the dialectical insuperability of form and content in the writings of the Frankfurt School, especially those of Adorno. See, for instance, Weber, "Translating the Untranslatable," 11.

88. Kracauer, "On Jewish Culture," 56–57. In 1926, Kracauer makes a similar comment regarding Franz Kafka (see K 5.2:494). On Kracauer's relationship to Judaism and Jewish identity, see Handelman, "Dialectics of Otherness."

89. Adorno, *Gesammelte Schriften*, 10.1:29.

90. Kracauer, *History*, 123.

91. Habermas, "Entwinement of Myth and Enlightenment," 119.

92. Kracauer, *History*, 62.

CONCLUSION

1. Liu, "Where Is Cultural Criticism in the Digital Humanities?"

2. Horkheimer, *Gesammelte Schriften*, 4:165.

3. See J. H. Plumb, *Crisis in the Humanities*. In 2010 and again in 2013, periodicals such as the *Atlantic*, the *New York Times*, and the *Wall Street Journal* published articles describing how, even at elite intuitions, the humanities are "attracting fewer undergraduates amid concerns about the degree's value in a rapidly changing job market," Levitz and Belkin, "Humanities Fall from Favor." See also Lewin, "As Interest Fades in the Humanities, Colleges Worry"; and Greteman, "It's the End of the Humanities as We Know It." Others have taken to defending the humanities as a place for innovation essential to STEM disciplines; see, for example, Edelstein, "How Is Innovation Taught?"

4. The origins of "humanities computing" are widely attributed to the work of an Italian priest, Roberto Busa, who in 1949 began using computers to make an *index verborum* for the works of Thomas Aquinas. See Hockey, "History of Humanities Computing," 4. On new German media theory and the digital, see Kittler, "Computeranalphabetismus" and "Es gibt keine Software." See also Koepnick and McGlothlin, *After the Digital Divide?*

5. On practices and debates in the digital humanities, see Burdick et al., *Digital_Humanities*; Schreibman, Siemens, and Unsworth, *New Companion to Digital Humanities*; and Gold and Klein, *Debates in the Digital Humanities 2016*.

6. Here I refer to the phrase of Rita Raley quoted in Kirschenbaum, "What Is 'Digital Humanities'?," 47–48. For Kirschenbaum, as a "discursive construction" the term *digital humanities* only slightly resembles the "material conduct" of humanists who employ digital tools; he lists the "terrible things" detractors have said about digital humanities (50).

7. On the similarities and differences of these computational approaches to culture, see Moretti, *Distant Reading*; Jockers, *Macroanalysis*; and Piper, "There Will Be Numbers." On the lack of diverse voices in these approaches,

see Nowviskie, "What Do Girls Dig?"; and Rhody, "Why I Dig: Feminist Approaches to Text Analysis."

8. Moretti, "Conjectures on World Literature," 57. Here I refer to the work of Rhody, "Topic Modeling and Figurative Language"; and Long and So, "Literary Pattern Recognition."

9. Inability to "read" all published texts is a common justification for the venture of distant reading, see Moretti, "Conjectures on World Literature," 55. Matt Jockers uses a similar justification, citing the Internet Archive and the academic research branch of Google Books, HathiTrust; see Jockers, *Macroanalysis*, 7–8. In German Studies, see Erlin and Tatlock, "Introduction," 1–2.

10. Kirsch, "Technology Is Taking Over English Departments." See a similar argument made in Brennan, "Digital-Humanities Bust."

11. As these responses point out, Kirsch's criticism, most prominently, mischaracterizes the debates over the definition of the digital humanities, obscures women's voices in the field, and misrepresents the digital humanities' own grasp of their place in the history of technology and interpretation. See, for instance, the reply of Burdick et al., "Immense Promise of the Digital Humanities." For other responses, see Sample, "Difficult Thinking about the Digital Humanities," 510; and Cordell, "On Ignoring Encoding."

12. Lotze, *Logik*, 259–260.

13. Adorno and Horkheimer, *Briefe und Briefwechsel*, 4.1:195.

14. Moretti, *Graphs, Maps, Trees*, 1. Suspicious of criticism's purported "distance" to its cultural object, scholars in cultural analytics such as Andrew Piper argue that qualities such as Moretti's particular notion of "distance" (for example, "reductiveness," "scale," and "generality") can "serve as the very conditions of enabling arguments to circulate more widely and, potentially, to create social change"; see Piper, "There Will Be Numbers."

15. Jockers, *Macroanalysis*, 37, 44, and 47. Jockers argues against Charles Fanning's claim of a "lost generation," which Fanning makes in *The Irish Voice in America* (2000): "It appears that [Fanning's] lost generation is in fact a lost generation of eastern, and probably male, Irish Americans with a penchant for writing about urban themes" (44).

16. Klein, "Image of Absence." See also Risam, "Beyond the Margins"; and Gil's work in Minimal Computing.

17. Allington, Brouillette, and Golumbia, "Neoliberal Tools (and Archives)."

18. Liu, "Where Is Cultural Criticism in the Digital Humanities?," 491. Examples of such media-theoretical work can be found in Galloway, *Protocol*; and Wark, *A Hacker Manifesto*.

19. Risam, "Digital Humanities in Other Contexts."

20. See "Omeka," which is developed by the Roy Rosenzweig Center for History and New Media at George Mason University and the Corporation for Digital Scholarship. Minimal Computing is a working group at GO::DH (Global Outlook::Digital Humanities), which is a Special Interest Group of the Alliance of Digital Humanities Organisations (ADHO). See Gil and Ortega, "Global Outlook"; and Gil, "User, the Learner and the Machines We Make."

21. See, most distinctively, the language and ideas employed in Schnapp, "Digital Humanities Manifesto 2.0."

22. Burdick et al., *Digital_Humanities*, vii. See also Erlin, "Digital Humanities Masterplots."

23. Burdick et al., *Digital_Humanities*, 122.

24. On the digital humanities' links to business and science, see ibid., 12, 18, 42. For criticism of the alignment of the digital humanities, the corporate university, and the sciences see Allington, Brouillette, and Golumbia, "Neoliberal Tools (and Archives)"; Allington "Managerial Humanities" (including the discussion in the comments); and Brennan, "Digital-Humanities Bust," in which the digital humanities' links with science and the corporatization of the university serve as "the obvious culprits" for all the "digital-humanities excitement."

25. Allington, Brouillette, and Golumbia, "Neoliberal Tools (and Archives)."

26. Ibid.

27. Ibid. David Golumbia makes a similar point in *The Cultural Logic of Computation*, 3. As Chad Wellmon points out, humanists have been in the business of collecting and evaluating data since the late nineteenth century; see Wellmon, "Loyal Workers and Distinguished Scholars."

28. Adorno and Horkheimer, *Briefe und Briefwechsel*, 4.1:195 and 255.

29. Horkheimer, *Gesammelte Schriften*, 4:140.

30. Berry, *Critical Theory and the Digital*, 19. I agree with Berry that we find an argument similar to Horkheimer and Adorno in not only Allington, Brouillette, and Golumbia's "Neoliberal Tools (and Archives)" but also Golumbia's *Cultural Logic of Computation*—neither of which mention *Dialectic of Enlightenment*.

31. Golumbia, *Cultural Logic of Computation*, 5.

32. I would like to thank Laura Portwood-Stacer for pointing out the resonances of Kracauer's position here with Patricia Hill Collins's concept of the "outsider-within" (see *Black Feminist Thought*, 10–11) and W. E. B. Du Bois's concept of "double-consciousness" (see *Souls of Black Folk*, 5–6).

33. Kracauer, *Werke*, 5.2:623. See similar calls to work and think "through" computation, as in Galloway, *Protocol*, 17; and Wark, *Hacker Manifesto*.

34. Posner, "What's Next," 35. One example of such a project would be Minimal Computing, mentioned earlier. It promotes accessibility by developing computational and digital infrastructures that do "not overburden the systems and bandwidth available to our colleagues around the world"; Gil and Ortega, "Global Outlooks in Digital Humanities," 31.

35. Jockers, *Macroanalysis*, 171.

36. Chun, "Working the Digital Humanities," 5. See also Cecire, "Introduction."

37. Scholars have charted theses "futures" of critical theory. For instance, Peter Uwe Hohendahl traces the past discourse of rationality from the first through second generations of critical theory; see Hohendahl, "From the Eclipse of Reason," 18–26, and other contributions in Fischer and Hohendahl, *Critical Theory*. Another future for the project of critical theory lies in "constitutive negativity" as a "practice of meaning destruction and renewal"; see Bernstein, *Recovering Ethical Life*, 234.

Abel, Carl. *Über den Gegensinn der Urworte.* Leipzig: Wilhelm Friedrich, 1884.

Adorno, Theodor W. *Aesthetic Theory.* London: Bloomsbury, 2013.

———. *Gesammelte Schriften.* Vols. 1–20. Edited by Rolf Tiedemann et al. Frankfurt am Main: Suhrkamp, 1970–1986.

Adorno, Theodor W., and Walter Benjamin. *Briefe und Briefwechsel.* Vol. 1. Edited by Henri Lonitz. Frankfurt am Main: Suhrkamp, 1994.

Adorno, Theodor W., and Max Horkheimer. *Briefe und Briefwechsel.* Vol. 4.1. Edited by Christoph Gödde and Henri Lonitz. Frankfurt am Main: Suhrkamp, 2003.

Adorno, Theodor W., and Gershom Scholem. *"Der liebe Gott wohnt im Detail" Briefwechsel 1939–1969.* Edited by Asaf Angermann. Frankfurt am Main: Suhrkamp, 2015.

Albrecht, Andrea. "'[H]eute gerade nicht mehr aktuell': Käte Hamburgers Novalis-Deutung im Kontext des Marburger Neukantianismus und der deutschen Geistesgeschichte." In *Käte Hamburger: Kontext, Theorie und Praxis,* edited by Claudia Löschner and Andrea Albrecht, 11–76. Berlin: De Gruyter, 2015.

Alexander, Amir. *Infinitesimal: How a Dangerous Mathematical Theory Shaped the Modern World.* New York: Farrar, Straus and Giroux, 2015.

Allen, Amy. *The End of Progress: Decolonizing the Normative Foundations of Critical Theory.* New York: Columbia University Press, 2016.

Allington, Daniel. "The Managerial Humanities; or, Why the Digital Humanities Don't Exist." Blogpost. March 31, 2013. Accessed August 23, 2018. http://www.danielallington.net/2013/03/the-managerial-humanities-or-why-the-digital-humanities-dont-exist/.

Allington, Daniel, Sarah Brouillette, and David Golumbia. "Neoliberal Tools (and Archives): A Political History of Digital Humanities." *Los Angeles*

Review of Books. Accessed January 17, 2017. https://lareviewofbooks.org /article/neoliberal-tools-archives-political-history-digital-humanities/.

Alter, Robert. *Necessary Angels: Tradition and Modernity in Kafka, Benjamin, and Scholem.* Cambridge, MA: Harvard University Press, 1991.

Altmann, Alexander. *Moses Mendelssohn: A Biographical Study.* Philadelphia: Jewish Publication Society of America, 1973.

Aquinas, Thomas. *Basic Works.* Edited by Jeffrey Hause and Robert Pasnau. Indianapolis: Hackett Publishing, 2014.

———. *Commentary on Aristotle's Physics.* Translated by Richard J. Blackwell, Richard J. Spath, and W. Edmund Thrilkel. Notre Dame, IN: Dumb Ox Books, 1999.

———. *Texte zum Gottesbeweis.* Edited by Engelbert Krebs. Bonn: Marcus und Weber, 1912.

Aristotle. *Categories. On Interpretation. Prior Analytics.* Translated by H. P. Cooke and Hugh Tredennick. Cambridge, MA: Harvard University Press, 1938.

———. *On the Heavens.* Translated by William Keith Chambers Guthrie. Cambridge, MA: Harvard University Press, 1939.

———. *Physics.* Translated by Robin Waterfield. Oxford: Oxford University Press, 1999.

———. *Posterior Analytics.* Translated by Hugh Tredennick. Cambridge, MA: Harvard University Press, 1960.

Ayer, A. J. "Editor's Introduction." In *Logical Positivism.* New York: Free Press, 1966.

Baggini, Julian, and Peter S. Fosl. *The Philosopher's Toolkit: A Compendium of Philosophical Concepts and Methods.* 2nd ed. Oxford: Wiley-Blackwell, 2010.

Barouch, Lina. "Lamenting Language Itself: Gershom Scholem on the Silent Language of Lamentation." *New German Critique* 37, no. 3 (2010): 1–26.

Batnitzky, Leora. *Idolatry and Representation: The Philosophy of Franz Rosenzweig Reconsidered.* Princeton, NJ: Princeton University Press, 2000.

Bauch, Bruno. "Lotzes Logik und ihre Bedeutung im deutschen Idealismus." *Beiträge zur Philosophie des deutschen Idealismus,* no. 1 (1919): 45–58.

Baumann, Stephanie. "Drei Briefe - Franz Rosenzweig an Siegfried Kracauer." *Zeitschrift für Religions- und Geistesgeschichte* 63, no. 2 (2011): 166–176.

Baumeister, Karl August. *Handbuch der Erziehungs- und Unterrichtslehre für höhere Schulen.* München: Beck, 1898.

Beiser, Frederick C. *The Fate of Reason: German Philosophy from Kant to Fichte.* Cambridge, MA: Harvard University Press, 2009.

———. *The German Historicist Tradition.* Oxford: Oxford University Press, 2011.

Belke, Ingrid. "Siegfried Kracauer: Geschichte einer Begegnung." In *Grenzgänge zwischen Dichtung, Philosophie und Kulturkritik: Über Margarete*

Susman, edited by Anke Gilleir and Barbara Hahn, 35–61. Göttingen: Wallstein Verlag, 2012.

Belke, Ingrid, and Irina Renz, eds. *Marbacher Magazin*. Vol. 47. Marbach am Neckar: Deutsche Schillergesellschaft, 1988.

Beller, Steven. *Vienna and the Jews, 1867–1938: A Cultural History*. Cambridge: Cambridge University Press, 1991.

Benacerraf, Paul, and Hilary Putnam, eds. *Philosophy of Mathematics: Selected Readings*. 2nd ed. Cambridge: Cambridge University Press, 1983.

Bendavid, Lazarus. *Versuch einer logischen Auseinandersetzung des Mathematischen Unendlichen*. Berlin: Petit und Schöne, 1789.

Benhabib, Seyla. *Critique, Norm, and Utopia: A Study of the Foundations of Critical Theory*. New York: Columbia University Press, 1986.

Benjamin, Mara. *Rosenzweig's Bible: Reinventing Scripture for Jewish Modernity*. Cambridge: Cambridge University Press, 2013.

Benjamin, Walter. *Gesammelte Briefe*. Vol. 1. Edited by Christoph Gödde. Frankfurt am Main: Suhrkamp, 1995.

———. *Gesammelte Schriften*. Vols. 1–7. Edited by Rolf Tiedemann and Hermann Schweppenhäuser. Frankfurt am Main: Suhrkamp, 1972–1991.

———. *The Origin of German Tragic Drama*. Translated by John Osborne. London: Verso, 2003.

Bergmann, Birgit. "Mathematics in Culture." In *Transcending Tradition: Jewish Mathematicians in German Speaking Academic Culture*, edited by Birgit Bergmann, Moritz Epple, and Ruti Ungar, 186–195. Berlin: Springer, 2012.

Bergmann, Hugo. "Maimon und Cohen." *Monatsschrift für Geschichte und Wissenschaft des Judentums* (1939): 548–561.

Bergson, Henri. *Creative Evolution*. Translated by Arthur Mitchell. New York: Henry Holt and Company, 1911.

Bernstein, Jay M. *Recovering Ethical Life: Jürgen Habermas and the Future of Critical Theory*. New York: Routledge, 2014.

Berry, David M. *Critical Theory and the Digital*. London: Bloomsbury, 2015.

Biale, David. *Gershom Scholem: Kabbalah and Counter-History*. 2nd ed. Cambridge, MA: Harvard University Press, 1982.

Blumenberg, Hans. *The Legitimacy of the Modern Age*. Cambridge, MA: MIT Press, 1985.

———. *Paradigms for a Metaphorology*. Translated by Robert Savage. Ithaca, NY: Cornell University Press, 2010.

Bomski, Franziska. *Die Mathematik im Denken und Dichten von Novalis: zum Verhältnis von Literatur und Wissen um 1800*. Berlin: De Gruyter, 2014.

Boole, George. *An Investigation of the Laws of Thought*. London: Walton and Maberly, 1854.

Boyer, Carl B. *History of Analytic Geometry*. New York: Dover Publications, 2004.

———. *The History of the Calculus and Its Conceptual Development*. New York: Dover Publications, 1959.

Boyer, Carl B., and Uta C. Merzbach. *A History of Mathematics*. 3rd ed. Hoboken, NJ: Wiley, 2011.

Bradshaw, David, Laura Marcus, and Rebecca Roach, eds. *Moving Modernisms: Motion, Technology, and Modernity*. Oxford: Oxford University Press, 2016.

Brasser, Martin, ed. *Rosenzweig als Leser: Kontextuelle Kommentare zum "Stern der Erlösung."* Berlin: De Gruyter, 2004.

Brennan, Timothy. "The Digital-Humanities Bust." *Chronicle of Higher Education*, October 15, 2017. Accessed August 23, 2018. https://www.chronicle.com/article/The-Digital-Humanities-Bust/241424.

Brenner, Michael. *The Renaissance of Jewish Culture in Weimar Germany*. New Haven, CT: Yale University Press, 1998.

Brüggemann, Heinz. *Das andere Fenster: Einblicke in Häuser und Menschen: zur Literaturgeschichte einer urbanen Wahrnehmungsform*. Frankfurt am Main: Fischer, 1989.

Brunkhorst, Hauke. *Adorno and Critical Theory*. Cardiff: University of Wales Press, 1999.

Bubner, Rüdiger. "Kann Theorie ästhetisch werden? Zum Hauptmotiv der Philosophie Adornos." In *Materialien zur ästhetischen Theorie Theodor W. Adornos: Konstruktion der Moderne*, edited by Burkhardt Lindner and W. Martin Lüdke, 108–137. Frankfurt am Main: Suhrkamp, 1985.

Buck-Morss, Susan. *The Origin of Negative Dialectics*. New York: Free Press, 1977.

Budde, Carl. "Das Hebräische Klagelied." *Zeitschrift für die alttestamentliche Wissenschaft* 1, no. 2 (1882): 1–52.

Burdick, Anne, Johanna Drucker, Peter Lunenfeld, Todd Presner, and Jeffery Schnapp. *Digital_Humanities*. Cambridge, MA: MIT Press, 2012.

———. "The Immense Promise of the Digital Humanities." *New Republic*, May 12, 2014. Accessed August 23, 2018. https://newrepublic.com/article/117711/digital-humanities-have-immense-promise-response-adam-kirsh.

Carnap, Rudolf. *Abriss der Logistik mit Besonderer Berücksichtigung der Relationstheorie und ihrer Anwendungen*. Wien: Springer, 1929.

———. "Die alte und die neue Logik." *Erkenntnis*, no. 1 (1930): 12–26.

———. "Die physikalische Sprache als Universalsprache der Wissenschaft." *Erkenntnis*, no. 2 (1931): 432–465.

———. *The Logical Structure of the World*. Translated by Rolf A. George. Chicago: Open Court Publishing, 1969.

———. "Über Protokollsätze." *Erkenntnis*, no. 3 (1932): 215–228.

———. *The Unity of Science.* Translated by Max Black. New York: Routledge, 2011.

———. "Wahrheit und Bewährung." *Actes du Congrès International de Philosophie Scientifique,* no. 4 (1936): 18–23.

Cassirer, Ernst. *The Philosophy of Symbolic Forms.* Vol. 3. Translated by Ralph Manheim. New Haven, CT: Yale University Press, 1957.

Cecire, Natalia. "Introduction: Theory and the Virtues of Digital Humanities." *Journal of Digital Humanities.* March 9, 2012. Accessed August 23, 2018. http://journalofdigitalhumanities.org/1-1/introduction-theory-and-the-virtues-of-digital-humanities-by-natalia-cecire/.

Cesàro, Ernesto. *Vorlesungen über natürliche Geometrie.* Translated by Gerhard Kowalewski. Leipzig: Teubner, 1901.

Chamberlain, Houston Stewart. *Immanuel Kant: Die Persönlichkeit als Einführung in das Werk.* 3rd ed. München: F. Bruckmann, 1916.

Chun, Wendy Hui Kyong. "Working the Digital Humanities: Uncovering Shadows between the Dark and the Light." Edited by Lisa Marie Rhody. *Differences* 25, no. 1 (2014): 1–25.

Cohen, Hermann. *Die Religion der Vernunft aus den Quellen des Judentums.* Leipzig: Fock, 1919.

———. *Werke.* Vols. 1–12. Edited by Helmut Holzhey et al. Hildesheim: G. Olms, 1977–.

Collins, Patricia Hill. *Black Feminist Thought: Knowledge, Consciousness, and the Politics of Empowerment.* 2nd ed. New York: Routledge, 2002.

Cook, Deborah. *Adorno on Nature.* New York: Routledge, 2014.

Cordell, Ryan. "On Ignoring Encoding." Accessed September 20, 2017. http://ryancordell.org/research/dh/on-ignoring-encoding/.

Craver, Harry T. *Reluctant Skeptic: Siegfried Kracauer and the Crises of Weimar Culture.* New York: Berghahn Books, 2017.

Dahms, Hans-Joachim. "The Emigration of the Vienna Circle." In *The Cultural Exodus from Austria: Vertreibung der Vernunft,* edited by Friedrich Stadler and Peter Weibel, 2nd ed., 57–79. Wien: Springer, 1995.

———. *Positivismusstreit: Die Auseinandersetzungen der Frankfurter Schule mit dem logischen Positivismus, dem amerikanischen Pragmatismus und dem kritischen Rationalismus.* Frankfurt am Main: Suhrkamp, 1994.

Dathe, Uwe. "Jena—Eine Episode aus Gershom Scholems Leben." *Zeitschrift für Religions- und Geistesgeschichte* 60, no. 1 (2008): 73–78.

Davis, Martin. *The Universal Computer: The Road from Leibniz to Turing.* 3rd ed. Boca Roton, FL: Taylor and Francis, 2018.

Dedekind, Richard. *Was sind und was sollen die zahlen.* 3rd ed. Braunschweig: Vieweg, 1911.

Derrida, Jacques. *Dissemination.* Translated by Barbara Johnson. Chicago: University of Chicago Press, 1983.

———. *Specters of Marx: The State of the Debt, the Work of Mourning and the New International.* Translated by Peggy Kamuf. New York: Routledge, 2012.

Descartes, René. *Discourse on Method, Optics, Geometry, and Meteorology.* Translated by Paul J. Olscamp. Indianapolis: Hackett Publishing, 2001.

Dimock, Wai Chee. *Through Other Continents: American Literature across Deep Time.* Princeton, NJ: Princeton University Press, 2008.

Du Bois, W. E. B. *The Souls of Black Folk.* London: Penguin, 1996.

Dubbels, Elke. *Figuren des Messianischen in Schriften deutsch-jüdischer Intellektueller 1900–1933.* Berlin: De Gruyter, 2011.

Dubiel, Helmut. *Theory and Politics: Studies in the Development of Critical Theory.* Cambridge, MA: MIT Press, 1985.

Duffy, Simon. "Maimon's Theory of Differentials as the Elements of Intuitions." *International Journal of Philosophical Studies* 22, no. 1 (2014): 1–20.

Düttmann, Alexander García. *Philosophy of Exaggeration.* Translated by James Philips. London: Continuum, 2007.

Edelstein, Dan. "How Is Innovation Taught? On the Humanities and the Knowledge Economy." *Liberal Education* 96, no. 1 (2010): 14–19.

Eiland, Howard, and Michael W. Jennings. *Walter Benjamin: A Critical Life.* Cambridge, MA: Harvard University Press, 2014.

Engel, Amir. *Gershom Scholem: An Intellectual Biography.* Chicago: University of Chicago Press, 2017.

Erlin, Matt, and Lynne Tatlock. "Introduction: 'Distant Reading' and the Historiography of Nineteenth-Century German Literature." In *Distant Readings: Topologies of German Culture in the Long Nineteenth Century,* edited by Matt Erlin and Lynne Tatlock. Rochester, NY: Camden House, 2014.

Euclid. *The Thirteen Books of Euclid's Elements.* Vol 1. Translated by Thomas L. Heath. New York: Dover Publications, 1956.

Fanning, Charles. *The Irish Voice in America: Irish-American Fiction from the 1760s to the 1980s.* Lexington, KY: University Press of Kentucky, 1990.

Feenberg, Andrew. *Critical Theory of Technology.* Oxford: Oxford University Press, 1991.

———. *The Philosophy of Praxis: Marx, Lukács and the Frankfurt School.* London: Verso, 2014.

Fenves, Peter. *Arresting Language: From Leibniz to Benjamin.* Stanford, CA: Stanford University Press, 2002.

———. *The Messianic Reduction: Walter Benjamin and the Shape of Time.* Stanford, CA: Stanford University Press, 2010.

Ferber, Ilit. "A Language of the Border: On Scholem's Theory of Lament." *Journal of Jewish Thought and Philosophy* 21, no. 2 (2013): 161–186.

Ferber, Ilit, and Paula Schwebel, eds. *Lament in Jewish Thought, Philosophical, Theological, and Literary Perspectives.* Berlin: De Gruyter, 2014.

Fichte, J. G. *The Science of Knowledge: With the First and Second Introductions.* Translated by Peter Heath and John Lachs. Cambridge: Cambridge University Press, 1970.

Field, Geoffrey G. *Evangelist of Race: The Germanic Vision of Houston Stewart Chamberlain.* New York: Columbia University Press, 1981.

Field, Hartry H. *Realism, Mathematics, and Modality.* Oxford: Blackwell, 1991.

Fiorato, Pierfrancesco, and Hartwig Wiedebach. "Hermann Cohen im Stern der Erlösung." In *Rosenzweig als Leser: Kontextuelle Kommentare zum "Stern der Erlösung,"* edited by Martin Brasser, 305–355. Tübingen: Niemeyer Verlag, 2004.

Fisher, Jaimey, and Peter-Uwe Hohendahl, eds. *Critical Theory: Current State and Future Prospects.* New York: Berghahn Books, 2001.

Foucault, Michel. *The Archaeology of Knowledge and the Discourse on Language.* New York: Vintage, 1982.

Frege, Gottlob. *Begriffsschrift.* 2nd ed. Hildesheim: Georg Olms Verlag, 1993.

———. *The Foundations of Arithmetic: A Logico-Mathematical Enquiry into the Concept of Number.* Translated by J. L. Austin. Evanston, IL: Northwestern University Press, 1980.

Freud, Sigmund. *The Standard Edition of the Complete Psychological Works of Sigmund Freud.* Vol. 12. Translated by James Strachey. London: Hogarth Press, 1958.

Friedman, Michael. *A Parting of the Ways: Carnap, Cassirer, and Heidegger.* Chicago: Open Court, 2000.

———. *Kant and the Exact Sciences.* Cambridge, MA: Harvard University Press, 1992.

———. *Reconsidering Logical Positivism.* Cambridge: Cambridge University Press, 1999.

Frisby, David. *Fragments of Modernity: Theories of Modernity in the Work of Simmel, Kracauer and Benjamin.* New York: Routledge, 2013.

———. "Zwischen den Sphären: Siegfried Kracauer und der Detektivroman." In *Siegfried Kracauer: Neue Interpretationen,* edited by Thomas Y. Levin and Michael Kessler, 39–58. Tübingen: Stauffenberg, 1990.

Funkenstein, Amos. *Perceptions of Jewish History.* Berkeley: University of California Press, 1993.

Galison, Peter. "Aufbau/Bauhaus: Logical Positivism and Architectural Modernism." *Critical Inquiry* 16, no. 4 (1990): 709–752.

Galloway, Alexander R. *Protocol: How Control Exists After Decentralization.* Cambridge, MA: MIT Press, 2004.

Gamwell, Lynn. *Mathematics and Art: A Cultural History.* Princeton, NJ: Princeton University Press, 2015.

Gemünden, Gerd, and Johannes von Moltke. "Introduction: Kracauer's Legacies." In *Culture in the Anteroom*, 1–29. Ann Arbor: University of Michigan Press, 2012.

Geulen, Eva. "Mega Melancholia: Adorno's *Minima Moralia.*" In *Critical Theory: Current State and Future Prospects*, edited by Jaimey Fisher and Peter-Uwe Hohendahl, 49–68. New York: Berghahn Books, 2001.

Gibbs, Robert. *Correlations in Rosenzweig and Levinas.* Princeton, NJ: Princeton University Press, 1994.

Gil, Alexander. "The User, the Learner and the Machines We Make." Minimal Computing, May 21, 2015. Accessed August 23, 2018. http://go-dh.github.io/mincomp/thoughts/2015/05/21/user-vs-learner/.

Gil, Alexander, and Élika Ortega. "Global Outlooks in Digital Humanities: Multilingual Practices and Minimal Computing." In *Doing Digital Humanities: Practice, Training, Research*, edited by Constance Crompton, Richard J. Lane, and Raymond George Siemens, 22–34. New York: Routledge, 2016.

Gilloch, Graeme. *Siegfried Kracauer.* Cambridge: Wiley, 2015.

Glatzer, Nahum Norbert. "The Frankfort Lehrhaus." *Leo Baeck Institute Yearbook*, no. 1 (1956): 105–122.

———. *Franz Rosenzweig: His Life and Thought.* Indianapolis: Hackett Publishing, 1998.

Global Outlook::Digital Humanities. Accessed February 13, 2018. http://www.globaloutlookdh.org/.

Gödel, Kurt. "Some Basic Theorems on the Foundations of Mathematics and Their Implications." In *Collected Works*, 3:290–323. Oxford: Oxford University Press, 1995.

Goethe, Johann Wolfgang. *Zur Farbenlehre.* Edited by Manfred Wenzel. In *Sämtliche Werke*, vol. 23. Frankfurt am Main: Deutsche Klassiker Verlag, 1991.

Goetschel, Willi. *The Discipline of Philosophy and the Invention of Modern Jewish Thought.* New York: Fordham University Press, 2013.

Gold, Matthew K., and Lauren F. Klein, eds. *Debates in the Digital Humanities 2016.* Minneapolis: University of Minnesota Press, 2016.

Golumbia, David. *The Cultural Logic of Computation.* Cambridge, MA: Harvard University Press, 2009.

Goodstein, Elizabeth. *Georg Simmel and the Disciplinary Imaginary.* Stanford, CA: Stanford University Press, 2017.

Gordon, Peter E. *Adorno and Existence*. Cambridge, MA: Harvard University Press, 2016.

———. *Continental Divide: Heidegger, Cassirer, Davos*. Cambridge, MA: Harvard University Press, 2012.

———. *Rosenzweig and Heidegger: Between Judaism and German Philosophy*. Berkeley: University of California Press, 2003.

———. "Science, Finitude, and Infinity: Neo-Kantianism and the Birth of Existentialism." *Jewish Social Studies* 6, no. 1 (1999): 30–53.

Grattan-Guinness, Ivor. "The Mathematical Turn in Logic." In *The Rise of Modern Logic: From Leibniz to Frege*, edited by Dov M. Gabbay and John Woods, 545–556. Amsterdam: Elsevier, 2004.

———. *The Search for Mathematical Roots, 1870–1940*. Princeton, NJ: Princeton University Press, 2000.

Greteman, Blaine. "It's the End of the Humanities as We Know It." *New Republic*, June 13, 2014. https://newrepublic.com/article/118139/crisis-humanities-has-long-history.

Guicciardini, Niccolò. "Newton's Method and Leibniz's Calculus." In *A History of Analysis*, edited by Hans Niels Jahnke, 73–103. Providence, RI: American Mathematical Society, 2003.

Gutzmer, August. *Reformvorschläge für den mathematischen und naturwissenschaftlichen Unterricht. Entworfen von der Unterrichtskommission der Gesellschaft deutscher Naturforscher und Ärzte*. Leipzig: Teubner, 1905.

Guyer, Paul. *Kant*. New York: Routledge, 2006.

Habermas, Jürgen. "The Entwinement of Myth and Enlightenment: Max Horheimer and Theodor Adorno." In *The Philosophical Discourse of Modernity. Twelve Lectures*, 106–130. Cambridge, MA: MIT Press, 1990.

———. *Knowledge and Human Interests*. Translated by Jeremy J. Shapiro. Boston, MA: Beacon Press, 1972.

———. "Modernity: An Unfinished Project." In *Habermas and the Unfinished Project of Modernity: Critical Essays on The Philosophical Discourse of Modernity*, edited by Maurizio Passerin d'Entrèves and Seyla Benhabib, 38–55. Cambridge, MA: MIT Press, 1997.

———. *The Theory of Communicative Action*. Vol. 1. Translated by Thomas McCarthy. Boston, MA: Beacon Press, 1985.

Hahn, Hans. "The Crisis in Intuition." In *Empiricism, Logic and Mathematics: Philosophical Papers*, edited by Brian McGuinness, 73–102. Berlin: Springer, 1980.

Hahn, Hans, Otto Neurath, and Rudolf Carnap. *Wissenschaftliche Weltauffassung. Der Wiener Kreis*. Wien: Arthur Wolf, 1929.

Handelman, Matthew. "The Dialectics of Otherness: Siegfried Kracauer's Figurations of the Jew, Judaism and Jewishness." *Yearbook for European Jewish Literature Studies* 2, no. 1 (2015): 90–111.

———. "The Forgotten Conversation. Five Letters from Franz Rosenzweig to Siegfried Kracauer, 1921–1923." *Scientia Poetica*, no. 15 (2011): 234–251.

Hanfling, Oswald. "Logical Positivism." In *Philosophy of Science, Logic and Mathematics in the Twentieth Century*, edited by Stuart G. Shanker, 193–213. London: Routledge, 1996.

Hansen, Miriam. *Cinema and Experience: Siegfried Kracauer, Walter Benjamin, and Theodor W. Adorno*. Berkeley: University of California Press, 2011.

———. "Decentric Perspectives: Kracauer's Early Writings on Film and Mass Culture." *New German Critique*, no. 54 (1991): 47–76.

———. "'With Skin and Hair': Kracauer's Theory of Film, Marseille 1940." *Critical Inquiry* 19, no. 3 (1993): 437–469.

Hegel, Georg Wilhelm Friedrich *Phenomenology of Spirit*. Translated by A. V. Miller. Oxford: Oxford University Press, 1977.

———. *The Science of Logic*. Translated by George di Giovanni. Cambridge: Cambridge University Press, 2010.

Heintz, Bettina. *Die Innenwelt der Mathematik: Zur Kultur und Praxis einer beweisenden Disziplin*. Berlin: Springer, 1999.

Hess, Jonathan M. *Germans, Jews and the Claims of Modernity*. New Haven, CT: Yale University Press, 2002.

Hewitt, Andrew. *Social Choreography: Ideology as Performance in Dance and Everyday Movement*. Durham, NC: Duke University Press, 2005.

Hewitt, Andrew. "Stumbling into Modernity: Body and Soma in Adorno." In *Critical Theory: Current State and Future Prospects*, edited by Jaimey Fisher and Peter-Uwe Hohendahl, 69–93. New York: Berghahn Books, 2001.

Hilbert, David. "Axiomatisches Denken." *Mathematische Annalen*, no. 78 (1917): 405–415.

Hockey, Susan. "The History of Humanities Computing." In *A Companion to Digital Humanities*, edited by Susan Schreibman, Ray Siemens, and John Unsworth. Malden, MA: Wiley-Blackwell, 2008.

Hohendahl, Peter Uwe. *The Fleeting Promise of Art: Adorno's Aesthetic Theory Revisited*. Ithaca, NY: Cornell University Press, 2013.

———. "From the Eclipse of Reason to Communicative Rationality and Beyond." In *Critical Theory: Current State and Future Prospects*, edited by Jaimey Fisher and Peter-Uwe Hohendahl, 3–28. New York: Berghahn Books, 2001.

Hollander, Dana. *Exemplarity and Chosenness: Rosenzweig and Derrida on the Nation of Philosophy*. Stanford, CA: Stanford University Press, 2008.

Honneth, Axel. *Pathologies of Reason: On the Legacy of Critical Theory.* New York: Columbia University Press, 2009.

Horkheimer, Max. *Critique of Instrumental Reason.* Translated by Matthew O'Connell. London: Verso, 2013.

———. *Eclipse of Reason.* London: Bloomsbury, 2013.

———. *Gesammelte Schriften.* Vols. 1–19. Edited by Alfred Schmidt and Gunzelin Schmid Noerr. Frankfurt am Main: Fischer, 1985–1996.

Husserl, Edmund. "Die Krisis der europäischen Wissenschaften und die transzendentale Phänomenologie. Eine Einleitung in die phänomenologische Philosophie." *Philosophia,* no. 1 (1936): 77–176.

———. *Ideas: General Introduction to Pure Phenomenology.* New York: Routledge, 2012.

———. "Philosophie als strenge Wissenschaft." In *Husserliana: Gesammelte Werke,* edited by Thomas Nenon and Hans Reiner Sepp, 25:3–62. Den Haag: M. Nijhoff, 1986.

———. *The Crisis of European Sciences and Transcendental Phenomenology: An Introduction to Phenomenological Philosophy.* Translated by David Carr. Evanston, IL: Northwestern University Press, 1970.

Huyssen, Andreas. *Miniature Metropolis: Literature in an Age of Photography and Film.* Cambridge, MA: Harvard University Press, 2015.

Idel, Moshe. *Kabbalah: New Perspectives.* New Haven, CT: Yale University Press, 1990.

———. *Old Worlds, New Mirrors: On Jewish Mysticism and Twentieth-Century Thought.* Philadelphia: University of Pennsylvania Press, 2010.

———. *Studies in Ecstatic Kabbalah.* Albany: State University of New York Press, 2012.

Jacobs, Jack. *The Frankfurt School, Jewish Lives, and Antisemitism.* Cambridge: Cambridge University Press, 2014.

Jacobson, Eric. *Metaphysics of the Profane: The Political Theology of Walter Benjamin and Gershom Scholem.* New York: Columbia University Press, 2010.

Jahnke, Hans Niels, ed. *A History of Analysis.* Providence: American Mathematical Society, 2003.

Janik, Allan, and Stephen Toulmin. *Wittgenstein's Vienna.* New York: Simon and Schuster, 1973.

Jay, Martin. *The Dialectical Imagination: A History of the Frankfurt School and the Institute of Social Research, 1923–1950.* Berkeley: University of California Press, 1973.

———. "The Extraterritorial Life of Siegfried Kracauer." In *Permanent Exiles: Essays on the Intellectual Migration from Germany to America,* 152–197. New York: Columbia University Press, 1986.

———. "The Politics of Translation: Siegfried Kracauer and Walter Benjamin on the Buber–Rosenzweig Bible." *Leo Baeck Institute Yearbook*, no. 21 (1976): 3–24.

Jockers, Matthew L. *Macroanalysis: Digital Methods and Literary History.* Urbana: University of Illinois Press, 2013.

Juola, Patrick, and Stephen Ramsay. *Six Septembers: Mathematics for the Humanist.* Lincoln, NE: Zea Books, 2017. Accessed August 23, 2018. http://digitalcommons.unl.edu/zeabook/55.

Kant, Immanuel. *Critique of Pure Reason.* Translated by Paul Guyer and Allen W. Wood. Cambridge: Cambridge University Press, 1999.

Kavka, Martin. *Jewish Messianism and the History of Philosophy.* Cambridge: Cambridge University Press, 2004.

———. "Verification (Bewährung) in Franz Rosenzweig." In *German-Jewish Thought Between Religion and Politics: Festschrift in Honor of Paul Mendes-Flohr on the Occasion of His Seventieth Birthday*, edited by Christian Wiese and Martina Urban, 167–184. Berlin: De Gruyter, 2012.

Kierkegaard, Søren. *Concluding Unscientific Postscript.* Translated by Alastair Hannay. Cambridge: Cambridge University Press, 2009.

Kirsch, Adam. "Technology Is Taking Over English Departments." *New Republic*, May 2, 2014. Accessed August 23, 2018. https://newrepublic.com/article/117428/limits-digital-humanities-adam-kirsch.

Kirschenbaum, Matthew. "What Is 'Digital Humanities,' and Why Are They Saying Such Terrible Things About It?" *Differences: A Journal of Feminist Cultural Studies* 25, no. 1 (2014): 46–63.

Kittel, Rudolf, ed. *Biblia Hebraica.* Leipzig: Hindrichs, 1913.

Kittler, Friedrich. "Computeranalphabetismus." In *Literatur im Informationszeitalter*, edited by Dirk Matejovski and Friedrich Kittler, 237–251. Frankfurt am Main: Campus-Verlag, 1996.

———. "Es gibt keine Software." In *Draculas Vermächtnis*, 225–242. Leipzig: Reclam, 1993.

Klein, Felix. *Vorträge über den mathematischen Unterricht an den Höheren Schulen.* Teil 1: *Von der Organisation des mathematischen Unterrichts.* Edited by Rudolf Schimmack. Leipzig: Teubner, 1907.

Klein, Lauren F. "The Image of Absence: Archival Silence, Data Visualization, and James Hemings." *American Literature* 85, no. 4 (2013): 661–688.

Koblitz, Neal. "Why STEM Majors Need the Humanities." *Chronicle of Higher Education*, January 6, 2017. Accessed August 23, 2018. http://www.chronicle.com/article/Why-STEM-Majors-Need-the/238833.

Koch, Gertrud "'Not Yet Accepted Anywhere': Exile, Memory, and Image in Kracauer's Conception of History." Translated by Jeremy Gaines. *New German Critique*, no. 54 (1991): 95–109.

———. *Siegfried Kracauer: An Introduction*. Translated by Jeremy Gaines. Princeton, NJ: Princeton University Press, 2000.

Koepnick, Lutz Peter, and Erin McGlothlin, eds. *After the Digital Divide? German Aesthetic Theory in the Age of New Media*. Rochester, NY: Camden House, 2009.

Kohnke, Klaus Christian. *The Rise of Neo-Kantianism: German Academic Philosophy between Idealism and Positivism*. Translated by R. J. Hollingdale. Cambridge: Cambridge University Press, 1991.

Kracauer, Siegfried. *History: The Last Things before the Last*. Princeton, NJ: Markus Wiener Publishers, 1995.

———. "On Jewish Culture." In *Siegfried Kracauer's American Writings: Essays on Film and Popular Culture*, edited by Johannes von Moltke and Kristy Rawson, 54–57. Berkeley: University of California Press, 2012.

———. *Werke*. Vols. 1–9. Edited by Inka Mülder-Bach and Ingrid Belke. Frankfurt am Main: Suhrkamp, 2006–2012.

Kracauer, Siegfried, and Theodor W. Adorno. *Briefwechsel 1923–1966*. Edited by Wolfgang Schopf. Frankfurt am Main: Suhrkamp, 2008.

Kracauer, Siegfried, and Walter Benjamin. *Briefe an Siegfried Kracauer: Mit vier Briefen von Siegfried Kracauer an Walter Benjamin*. Edited by Theodor W. Adorno Archiv. Marbach am Neckar: Deutsche Schillergesellschaft, 1987.

Kracauer, Siegfried, and Leo Löwenthal. *In steter Freundschaft: Briefwechsel Leo Löwenthal und Siegfried Kracauer 1922–1966*. Edited by Peter E. Jansen. Lüneburg: Klampen, 2003.

Kracauer, Siegfried, and Margarete Susman. Correspondence, 1920–1922. Susman Nachlass. Deutsches Literaturarchiv, Marbach am Neckar.

Kraus, Karl. "Heine und die Folgen." In *Schriften*, 4:185–210. Frankfurt am Main: Suhrkamp, 1989.

Kuntze, Friedrich. *Die Philosophie Salomon Maimons*. Heidelberg: C. Winter, 1912.

Kutschmann, Werner. *Der Naturwissenschaftler und sein Körper: die Rolle der "inneren Natur" in der experimentellen Naturwissenschaft der frühen Neuzeit*. Frankfurt am Main: Suhrkamp, 1986.

Lazier, Benjamin. *God Interrupted: Heresy and the European Imagination Between the World Wars*. Princeton, NJ: Princeton University Press, 2008.

Leibniz, Gottfried Wilhelm. "Letter to Varignon, with a Note on the 'Justification of the Infinitesimal Calculus by that of Ordinary Algebra' (1702)." In *Philosophical Papers and Letters*. Translated by Leroy E. Loemker. 2nd edition. Dordrecht: Kluwer Academic Publishers, 1969.

Levin, Thomas. *Siegfried Kracauer: Eine Bibliographie seiner Schriften*. Marbach am Neckar: Deutsche Schillergesellschaft, 1989.

———. "Introduction." In *The Mass Ornament: Weimar Essays*, edited by Thomas Y. Levin. Cambridge, MA: Harvard University Press, 1995.

Levitz, Jennifer, and Douglas Belkin. "Humanities Fall from Favor." *Wall Street Journal*, June 6, 2013. Accessed August 23, 2018. http://www.wsj.com/articles/SB10001424127887324069104578527642373232184.

Lewin, Tamar. "As Interest Fades in the Humanities, Colleges Worry." *New York Times*, October 30, 2013. Accessed August 23, 2018. http://www.nytimes.com/2013/10/31/education/as-interest-fades-in-the-humanities-colleges-worry.html.

Liu, Alan. "Where Is Cultural Criticism in the Digital Humanities?" In *Debates in the Digital Humanities 2012*, ed. Matthew K. Gold, 490–509. Minneapolis: University of Minnesota Press, 2012.

Lobenstein-Reichmann, Anja. *Houston Stewart Chamberlain—Zur textlichen Konstruktion einer Weltanschauung: Eine sprach-, diskurs- und ideologiegeschichtliche Analyse*. Berlin: De Gruyter, 2008.

Long, Hoyt, and Richard Jean So. "Literary Pattern Recognition: Modernism between Close Reading and Machine Learning." *Critical Inquiry* 42, no. 2 (2015): 235–267.

Lotze, Hermann. *Logik: Drei Bücher, vom Denken, vom Untersuchen und vom Erkennen*. 2nd ed. Leipzig: F. Meiner, 1912.

Lukács, Georg. *History and Class Consciousness: Studies in Marxist Dialectics*. Translated by Rodney Livingstone. Cambridge, MA: MIT Press, 1971.

Maimon, Salomon. *Essay on Transcendental Philosophy*. Translated by Nick Midgley, Henry Somers-Hall, Alistair Welchman, and Merten Reglitz. London: Continuum, 2010.

Maimon, Solomon. *Solomon Maimon: An Autobiography*. Translated by J. Clark Murray. Urbana: University of Illinois Press, 2001.

Marcuse, Herbert. *One-Dimensional Man: Studies in the Ideology of Advanced Industrial Society*. London: Routledge, 2013.

———. "Review of International Encyclopedia of Unified Science." *Zeitschrift für Sozialforschung*, no. 8 (1939): 228–232.

Marseille à la fin de l'Ancien Régime. Marseille: M. Laffitte, 1896.

Mauthner, Fritz. *Beiträge zu einer Kritik der Sprache.* 2nd ed. 3 vols. Stuttgart: Cotta, 1906.

Mehrtens, Herbert. *Moderne, Sprache, Mathematik.* Frankfurt am Main: Suhrkamp, 1990.

Mendelssohn, Moses. "On the Evidence in Metaphysical Sciences." In *Philosophical Writings*, translated by Daniel O. Dahlstrom. Cambridge: Cambridge University Press, 1997.

Mendes-Flohr, Paul R., and Jehuda Reinharz. "From Relativism to Religious Faith: The Testimony of Franz Rosenzweig's Unpublished Diaries." *Leo Baeck Institute Yearbook*, no. 22 (1977): 161–174.

Menninghaus, Winfried. *Walter Benjamins Theorie der Sprachmagie.* Frankfurt am Main: Suhrkamp, 1980.

Mertens, Bram. *Dark Images, Secret Hints: Benjamin, Scholem, Molitor and the Jewish Tradition.* Oxford: Peter Lang, 2007.

Minimal Computing. Accessed September 18, 2017. http://go-dh.github.io /mincomp/.

Molitor, Franz Joseph. *Philosophie der Geschichte oder über die Tradition.* Frankfurt am Main: Hermann'sche Buchhandlung, 1827.

Moltke, Johannes von. *The Curious Humanist: Siegfried Kracauer in America.* Berkeley: University of California Press, 2016.

———. "Teddie and Friedel: Theodor W. Adorno, Siegfried Kracauer, and the Erotics of Friendship." *Criticism* 51, no. 4 (2009): 683–694.

Monmonier, Mark. *Rhumb Lines and Map Wars: A Social History of the Mercator Projection.* Chicago: University of Chicago Press, 2010.

Moretti, Franco. "Conjectures on World Literature." *New Left Review* 2, no. 1 (2000): 54–68.

———. *Distant Reading.* London: Verso, 2013.

———. *Graphs, Maps, Trees: Abstract Models for Literary History.* London: Verso, 2007.

Mosès, Stéphane. *The Angel of History: Rosenzweig, Benjamin, Scholem.* Stanford, CA: Stanford University Press, 2008.

Mülder-Bach, Inka. "Nachbemerkung und editorische Notiz." In Kracauer, *Werke*, 1:375–392. Frankfurt am Main: Suhrkamp, 2006.

———. *Siegfried Kracauer: Grenzgänger zwischen Theorie und Literatur.* Stuttgart: Metzler, 1985.

Nernst, Walther, and Arthur Schönflies. *Einführung in die mathematische Behandlung der Naturwissenschaften. Kurzgefasstes Lehrbuch der Differential- und Integralrechnung mit besonderer Berücksichtigung der Chemie.* München: R. Oldenbourg, 1907.

Neurath, Otto. "Protokollsätze." *Erkenntnis*, no. 3 (1932): 204–214.

———. "Wissenschaftliche Weltauffasung. Der Wiener Kreis. [The Scientific Conception of the World. The Vienna Circle]." In *Empiricism and Sociology*, edited by Robert S. Cohen and Marie Neurath, 299–318. Dordrecht: Reidel, 1973.

Newton, Isaac. *Newtons Abhandlung über die Quadratur der Kurven (1704)*. Translated by Gerhard Kowalewski. Leipzig: Wilhelm Engelmann, 1908.

———. "The Principia." In *Isaac Newton: Philosophical Writings*, edited by Andrew Janiak, 40–93. Cambridge: Cambridge University Press, 2004.

Ng, Julia. "'+1': Scholem and the Paradoxes of the Infinite." *Rivista Italiana di Filosofia del Linguaggio* 8, no. 2 (2014): 196–210.

Nolan, Mary. *Visions of Modernity: American Business and the Modernization of Germany*. Oxford: Oxford University Press, 1994.

North, Paul. *The Yield: Kafka's Atheological Reformation*. Stanford, CA: Stanford University Press, 2015.

Novalis. *Novalis Schriften*. Vol. 2. Edited by Ludwig Tieck and Friedrich Schlegel. Berlin: G. Reimer, 1837.

Nowviskie, Bethany. "What Do Girls Dig?" In *Debates in the Digital Humanities*, edited by Matthew K. Gold, 235–240. Minneapolis: University of Minnesota Press, 2012.

Omeka. Accessed September 18, 2017. https://omeka.org/.

O'Connor, Brian. *Adorno*. New York: Routledge, 2013.

O'Neill, John, and Thomas Uebel. "Horkheimer and Neurath: Restarting a Disrupted Debate." *European Journal of Philosophy* 12, no. 1 (2004): 75–105.

Peckhaus, Volker. *Logik, Mathesis Universalis und Allgemeine Wissenschaft: Leibniz und die Wiederentdeckung der formalen Logik im 19. Jahrhundert*. Berlin: Akademie Verlag, 1997.

———. "The Mathematical Origins of Nineteenth-Century Algebra of Logic." In *The Development of Modern Logic*, edited by Leila Haaparanta, 159–195. Oxford: Oxford University Press, 2009.

Pines, Shlomo. "Der Islam im 'Stern der Erlösung,' Eine Untersuchung zu Tendenzen und Quellen Franz Rosenzweigs." *Hebräische Beiträge zur Wissenschaft des Judentums*, vols. 3–5 (1987–1989): 138–148.

Piper, Andrew. "There Will Be Numbers." *Journal of Cultural Analytics*, May 23, 2016. Accessed August 23, 2018. http://culturalanalytics.org/2016/05/there-will-be-numbers/.

Plato. *Meno and Other Dialogues*. Translated by Robin Waterfield. Oxford: Oxford University Press, 2009.

———. *Republic*. Edited by C. D. C. Reeve. Translated by G. M. A. Grube. 2nd edition. Indianapolis: Hackett Publishing, 1992.

Plumb, J. H., ed. *Crisis in the Humanities*. Harmondsworth: Penguin, 1964.

Poe, Edgar Allan. *Selected Tales*. Edited by David Van Leer. 2nd ed. Oxford: Oxford University Press, 2008.

Poincaré, Henri. *Wissenschaft und Hypothese*. Translated by F. and L. Lindemann. 3rd ed. Leipzig: Teubner, 1914.

Pollock, Benjamin. *Franz Rosenzweig and the Systematic Task of Philosophy*. Cambridge: Cambridge University Press, 2009.

———. *Franz Rosenzweig's Conversions: World Denial and World Redemption*. Bloomington: Indiana University Press, 2014.

Poma, Andrea. *The Critical Philosophy of Hermann Cohen*. Translated by John Denton. Albany: State University of New York Press, 1997.

Popper, Karl R. *Logik der Forschung*. In *Gesammelte Werke*, vol. 3, edited by Herbert Keuth. Tübingen: Mohr Siebeck, 2005.

Posner, Miriam. "What's Next: The Radical, Unrealized Potential of Digital Humanities." In *Debates in the Digital Humanities 2016*, edited by Matthew K. Gold and Lauren F. Klein, 33–41. Minneapolis: University of Minnesota Press, 2016.

Prochnik, George. *Stranger in a Strange Land: Searching for Gershom Scholem and Jerusalem*. New York: Other Press, 2017.

Pyenson, Lewis. *Neohumanism and the Persistence of Pure Mathematics in Wilhelmian Germany*. Philadelphia: American Philosophical Society, 1983.

Rabinbach, Anson. *In the Shadow of Catastrophe: German Intellectuals between Apocalypse and Enlightenment*. Berkeley: University of California Press, 2001.

———. *The Human Motor: Energy, Fatigue, and the Origins of Modernity*. Berkeley: University of California Press, 1992.

raca. "Zwei Flächen." *Frankfurter Zeitung*, September 26, 1926, Erstes Morgenblatt edition.

Rhody, Lisa M. "Topic Modeling and Figurative Language." *Journal of Digital Humanities* 2, no. 1 (2012). Accessed August 23, 2018. http://journalofdigitalhumanities.org/2-1/topic-modeling-and-figurative-language-by-lisa-m-rhody/.

———. "Why I Dig: Feminist Approaches to Text Analysis." In *Debates in Digital Humanities 2016*, edited by Matthew K. Gold and Lauren F. Klein, 536–540. Minneapolis: University of Minnesota Press, 2016.

Riecke, Eduard. *Lehrbuch der Experimental-Physik: Mechanik, Akustik, Optik*. Leipzig: Veit, 1896.

Risam, Roopika. "Beyond the Margins: Intersectionality and the Digital Humanities" 9, no. 2 (2015). Accessed August 23, 2018. http://www.digitalhumanities.org/dhq/vol/9/2/000208/000208.html.

———. "Digital Humanities in Other Contexts," May 3, 2016. Accessed August 23, 2018. http://roopikarisam.com/uncategorized/digital -humanities-in-other-contexts/.

Rosenzweig, Franz. "Abgangs-Zeugnis," 1905. Universitätsarchiv Göttingen. University of Göttingen.

———. *Die "Gritli"-Briefe: Briefe an Margrit Rosenstock-Huessy.* Edited by Inken Rühle and Reinhold Mayer. Tübingen: Bilam, 2002.

———. *Feldpostbriefe: Die Korrespondenz mit den Eltern.* Edited by Wolfgang D. Herzfeld. Freiburg: Verlag Karl Alber, 2013.

———. "'Forward' to Hegel and the State." In *Philosophical and Theological Writings*, translated by Paul Franks and Michael Morgan, 73–74. Indianapolis: Hackett Publishing, 2000.

———. *Hegel und der Staat.* Edited by Frank Lachmann. Berlin: Suhrkamp, 2010.

———. Letter to Siegfried Kracauer, May 25, 1923. In Stephanie Baumann. "Drei Briefe: Franz Rosenzweig an Siegfried Kracauer." *Zeitschrift für Religions- und Geistesgeschichte* 63, no. 2 (2011): 166–176.

———. "The New Thinking." In *Philosophical and Theological Writings*, translated by Paul Franks and Michael Morgan, 109–139. Indianapolis: Hackett Publishing, 2000.

———. *Philosophical and Theological Writings.* Translated by Paul Franks and Michael Morgan. Indianapolis: Hackett Publishing, 2000.

Russell, Bertrand. *The Principles of Mathematics.* Cambridge: Cambridge University Press, 1903.

Russell, Bertrand, and Alfred North Whitehead. *Principia Mathematica.* Cambridge: Cambridge University Press, 1910.

Salkowski, Erich. *Grundzüge der darstellenden Geometrie.* Leipzig: Akademische Verlagsgesellschaft, 1928.

Sample, Mark. "Difficult Thinking about the Digital Humanities." In *Debates in the Digital Humanities 2016*, edited by Matthew K. Gold and Lauren F. Klein, 510–513. Minneapolis: University of Minnesota Press, 2016.

Santner, Eric L. *On the Psychotheology of Everyday Life: Reflections on Freud and Rosenzweig.* Chicago: University of Chicago Press, 2007.

Sauter, Caroline. "Hebrew, Jewishness, and Love: Translation in Gershom Scholem's Early Work." *Naharaim* 9, nos. 1–2 (2015): 151–178.

Scheerbart, Paul. *Lesabéndio. Ein Asteroiden-Roman.* In *Gesammelte Werke*, vol. 5, edited by Thomas Bürk, Joachim Körber, and Uli Kohnle. Linkenheim: Phantasia, 1988.

Schivelbusch, Wolfgang. *Intellektuellendämmerung: Zur Lage der Frankfurter Intelligenz in den zwanziger Jahren.* Frankfurt am Main: Insel Verlag, 1982.

Schlüpmann, Heide. "The Subject of Survival: On Kracauer's Theory of Film." Translated by Jeremy Gaines. *New German Critique*, no. 54 (1991): 111–126.

Schmidt-Biggemann, Wilhelm. *Topica Universalis: Eine Modellgeschichte humanistischer und barocker Wissenschaft.* Hamburg: Meiner Verlag, 1983.

Schnapp, Jeffrey et al. "The Digital Humanities Manifesto 2.0." 2009. Accessed August 23, 2018. http://www.humanitiesblast.com/manifesto /Manifesto_V2.pdf.

Scholem, Gershom. "A List of the Books in Scholem's Library" (1923). Gershom Scholem Archive. National Library of Israel, Jerusalem.

———. "Against the Myth of the German-Jewish Dialogue." In *On Jews and Judaism in Crisis: Selected Essays*, edited by Werner Dannhauser, 61–64. New York: Schocken, 2012.

———. "Bezieht sich die reine Logik . . . ," 1917. Gershom Scholem Archive. National Library of Israel, Jerusalem.

———. *Briefe.* Vol. 1. Edited by Itta Shedletzky. München: C. H. Beck, 1994.

———. *Briefe an Werner Kraft.* Edited by Werner Kraft. Frankfurt am Main: Suhrkamp, 1986.

———. Gershom Scholem's Diaries, 1915–1917, Gershom Scholem Archive. National Library of Israel, Jerusalem.

———. *Lamentations of Youth: The Diaries of Gershom Scholem, 1913–1919.* Translated by Anthony David Skinner. Cambridge, MA: Harvard University Press, 2008.

———. *Major Trends in Jewish Mysticism.* New York: Schocken, 1995.

———. "The Messianic Idea in Judaism." In *The Messianic Idea in Judaism and Other Essays on Jewish Spirituality.* New York: Schocken, 1995.

———. "Ein mittelalterliches Klagelied." *Der Jude*, no. 4 (September 1919): 283–286.

———. "Der Name Gottes und die Sprachtheorie der Kabbala." In *Judaica 3*, 7–70. Frankfurt am Main: Suhrkamp, 1973.

———. "Notebook to Knopp's Lectures on Differential Equations." Gershom Scholem Archive. National Library of Israel, Jerusalem.

———. "On Lament and Lamentation." In *Lament in Jewish Thought*, edited by Paula Schwebel and Ilit Ferber, translated by Paula Schwebel and Lina Barouch, 313–319. Berlin: De Gruyter, 2014.

———. "On the 1930 Edition of Rosenzweig's Star of Redemption." In *The Messianic Idea in Judaism and Other Essays on Jewish Spirituality*, 320–324. New York: Schocken, 1995.

———. *On the Kabbalah and Its Symbolism.* New York: Schocken, 1996.

————. "On the Social Psychology of the Jews in Germany: 1900–1933." In *Jews and Germans from 1860 to 1933: The Problematic Symbiosis*, edited by David Bronsen, 9–32. Heidelberg: Winter, 1979.

————. *Tagebücher nebst Aufsätzen und Entwürfen bis 1923*. Vols. 1–2. Edited by Karlfried Gründer and Friedrich Niewöhner et al. Frankfurt am Main: Jüdischer Verlag, 1995–2000.

————. *Von Berlin nach Jerusalem. Jugenderinnerungen*. Frankfurt am Main: Suhrkamp, 1997.

————. *Walter Benjamin: Die Geschichte einer Freundschaft*. Frankfurt am Main: Suhrkamp, 1975.

————. "With Gershom Scholem: An Interview." In *On Jews and Judaism in Crisis: Selected Essays*, edited by Werner Dannhauser, 1–48. New York: Schocken, 1976.

————. "Zehn unhistorische Sätze über Kabbala." In *Judaica 3*, 264–271. Frankfurt am Main: Suhrkamp, 1973.

Scholem, Gershom, and Walter Benjamin. *Briefwechsel 1933–1940*. Frankfurt am Main: Suhrkamp, 1997.

Scholem, Gershom, and Werner Kraft. Correspondence, 1915–1971. Gershom Scholem Archive. National Library of Israel, Jerusalem.

Scholem, Gershom, and Greta Lissauer. Correspondence, 1916–1918. Gershom Scholem Archive. National Library of Israel, Jerusalem.

Schopenhauer, Arthur. *The World as Will and Representation*. Vol. 1. Translated by Christopher Janaway. Cambridge: Cambridge University Press, 2010.

Schreibman, Susan, Ray Siemens, and John Unsworth, eds. *A New Companion to Digital Humanities*. Malden, MA: Wiley, 2016.

Schröter, Michael. "Weltzerfall und Rekonstruktion: Zur Physiognomik Siegfried Kracauers." *Text + Kritik*, no. 68 (1980): 18–40.

Schubring, Gert. *Conflicts between Generalization, Rigor, and Intuition: Number Concepts Underlying the Development of Analysis in 17th–19th Century France and Germany*. Berlin: Springer, 2006.

————. "Mathematics Education in Germany (Modern Times)." In *Handbook on the History of Mathematics Education*, edited by Gert Schubring and Alexander Karp, 241–255. New York: Springer, 2014.

————. "Pure and Applied Mathematics in Divergent Institutional Settings in Germany: The Role and Impact of Felix Klein." In *The History of Modern Mathematics*, vol. 2: *Institutions and Applications*, edited by David E. Rowe and John McCleary, 171–220. Boston, MA: Academic Press, 1989.

Schudeiskÿ, Albert. *Projektionslehre: Die rechtwinklige Parallelprojektion und ihre Anwendung auf die Darstellung technischer Gebilde nebst einem Anhang über die schiefwinklige Parallelprojektion in kurzer, leicht fasslicher Behandlung für Selbstunterricht und Schulgebrauch.* Leipzig: Teubner, 1918.

Schulte, Christoph. "Messianism Without Messiah: Messianism, Religion, and Secularization in Modern Jewish Thought." In *Secularism in Question: Jews and Judaism in Modern Times,* edited by Ari Joskowicz and Ethan B. Katz. Philadelphia: University of Pennsylvania Press, 2015.

Schwebel, Paula. "The Tradition in Ruins: Walter Benjamin and Gershom Scholem on Language and Lament." In *Lament in Jewish Thought,* edited by Ilit Ferber and Paula Schwebel, 277–301. Berlin: De Gruyter, 2014.

Shakespeare, William. *Shakspeare's dramatische Werke.* Translated by August Wilhelm Schlegel. Berlin: Johann Friedrich Unger, 1798.

Silverman, Lisa. *Becoming Austrians: Jews and Culture between the World Wars.* Oxford: Oxford University Press, 2012.

Simmel, Georg. *The Philosophy of Money.* Translated by David Frisby. New York: Routledge, 2011.

Smith, David Woodruff. *Husserl.* New York: Routledge, 2013.

Smith, John H. "The Infinitesimal as Theological Principle: Representing the Paradoxes of God and Nothing in Cohen, Rosenzweig, Scholem, and Barth." *MLN* 127, no. 3 (2012): 562–588.

Später, Jörg. *Siegfried Kracauer: Eine Biographie.* Berlin: Suhrkamp, 2016.

Spinoza, Benedictus de. *Ethics.* Translated by Edwin M. Curley. London: Penguin Books, 2005.

Spector, Scott. "Modernism without Jews: A Counter-Historical Argument." *Modernism/Modernity* 13, no. 4 (2006): 615–633.

Stalder, Helmut. *Siegfried Kracauer: Das journalistische Werk.* Würzburg: Königshausen and Neumann, 2003.

Steiner, Rudolf. "Mathematik und Okkultismus." In *Rudolf Steiner Gesamtausgabe,* 2nd ed., 35:7–18. Dornach: Rudolf Steiner Verlag, 1984.

Susman, Margarete. "Der Exodus aus der Philosophie." *Frankfurter Zeitung,* June 17, 1921.

Thiel, Christian. *Grundlagenkrise und Grundlagenstreit: Studie über das normative Fundament der Wissenschaften am Beispiel von Mathematik und Sozialwissenschaft.* Meisenheim am Glan: Hain, 1972.

Tiedemann, Rolf. "Historical Materialism or Political Messianism? An Interpretation of the Theses 'On the Concept of History.'" In *The Frankfurt School: Critical Assessments,* edited by Jay Bernstein, 2:111–139. New York: Routledge, 1994.

Troeltsch, Ernst. *Der Historismus und seine Probleme.* Tübingen: C. B. Mohr, 1922.

Uebel, Thomas Ernst. *Empiricism at the Crossroads: The Vienna Circle's Protocol-Sentence Debate*. Chicago: Open Court, 2007.

Verzeichnis der Vorlesungen auf der Georg-August-Universität zu Göttingen während des Sommerhalbjahrs 1905. Göttingen: Dieterich'sche Universitäts-Buchdruckerei, W. Fr. Kaestner, 1905.

Vidler, Anthony. *Warped Space: Art, Architecture, and Anxiety in Modern Culture*. Cambridge, MA: MIT Press, 2002.

Voigts, Manfred. *Oskar Goldberg: Der mythische Experimentalwissenschaftler*. Berlin: Agora, 1992.

Volkert, Klaus Thomas. *Die Krise der Anschauung: Eine Studie zu formalen und heuristischen Verfahren in der Mathematik seit 1850*. Göttingen: Vandenhoeck und Ruprecht, 1986.

Voss, Aurel E. *Über das Wesen der Mathematik*. 2nd ed. Leipzig: Teubner, 1913.

Wark, McKenzie. *A Hacker Manifesto*. Cambridge, MA: Harvard University Press, 2009.

Waszek, Norbert. *Rosenzweigs Bibliothek: Der Katalog des Jahres 1939 mit einem Bericht über den derzeitigen Zustand in der tunesischen Nationalbibliothek*. Freiburg: Verlag Karl Alber, 2017.

Weatherby, Leif. *Transplanting the Metaphysical Organ: German Romanticism between Leibniz and Marx*. New York: Fordham University Press, 2016.

Weber, Max. "Die protestantischen Sekten und der Geist des Kapitalismus." In *Gesammelte Aufsätze zur Religionssoziologie*, 207–237. Tübingen: Mohr Siebeck, 1986.

———. "Wissenschaft als Beruf." In *Gesamtausgabe*, edited by Horst Baier, 1.17:71–111. Tübingen: Mohr Siebeck, 1992.

Weber, Samuel. *Benjamin's -abilities*. Cambridge, MA: Harvard University Press, 2010.

———. "Translating the Untranslatable." In *Prisms*. Cambridge, MA: MIT Press, 1983.

Weidner, Daniel. *Gershom Scholem. Politisches, esoterisches und historiographisches Schreiben*. München: Fink, 2003.

Weigel, Sigrid. "Scholems Gedichte und seine Dichtungstheorie: Klage, Adressierung, Gabe und das Problem einer biblischen Sprache in unserer Zeit." In *Gershom Scholem, Literatur und Rhetorik*, edited by Stéphane Mosès and Sigrid Weigel, 17–47. Köln: Böhlau, 2000.

Weissberg, Liliane. "Erfahrungsseelenkunde als Akkulturation: Philosophie, Wissenschaft und Lebensgeschichte bei Salomon Maimon." In *Der ganze Mensch: Anthropologie und Literaturwissenschaft im achtzehnten Jahrhundert*, edited by Hans Jürgen Schings, 298–328. Stuttgart: J. B. Metzler Verlag, 1994.

———. "Toleranzidee und Emanzipationsdebatte: Moses Mendelssohn, Salomon Maimon, Lazarus Bendavid." In *Toleranzdiskurse in der Frühen Neuzeit*, edited by Friedrich Vollhardt, 363–380. Berlin: De Gruyter, 2015.

———. *Über Haschisch und Kabbala: Gershom Scholem, Siegfried Unseld und das Werk von Walter Benjamin*. Marbach am Neckar: Deutsche Schillergesellschaft, 2012.

Weitz, Eric D. *Weimar Germany: Promise and Tragedy*. Princeton, NJ: Princeton University Press, 2013.

Wellmer, Albrecht. *Critical Theory of Society*. Translated by John Cumming. New York: Herder and Herder, 1971.

———. "Truth, Semblance, Reconciliation: Adorno's Aesthetic Redemption of Modernity." In *Frankfurt School: Critical Assessments*, edited by J. M. Bernstein, 4:29–54. London: Routledge, 1994.

Wellmon, Chad. "Loyal Workers and Distinguished Scholars: Big Humanities and the Ethics of Knowledge." *Modern Intellectual History* (2017): 1–39.

Wheatland, Thomas. *The Frankfurt School in Exile*. Minneapolis: University of Minnesota Press, 2009.

Wiggershaus, Rolf. *The Frankfurt School: Its History, Theories, and Political Significance*. Cambridge, MA: MIT Press, 1995.

———. *Wittgenstein und Adorno: Zwei Spielarten modernen Philosophierens*. Göttingen: Wallstein Verlag, 2000.

Winterhalter, Benjamin. "The Morbid Fascination with the Death of the Humanities." *Atlantic*, June 6, 2014. Accessed August 23, 2018. https://www.theatlantic.com/education/archive/2014/06/the-morbid-fascination-with-the-death-of-the-humanities/372216/.

Wittgenstein, Ludwig. *Tractatus Logico-Philosophicus*. Translated by D. F. Pears and B. F. McGuinness. 2nd ed. New York: Routledge, 2001.

Wolin, Richard. *Walter Benjamin: An Aesthetic of Redemption*. Berkeley: University of California Press, 1994.

Wolosky, Shira. "Gershom Scholem's Linguistic Theory." *Jerusalem Studies in Jewish Thought* 2 (2007): 165–205.

absence, 10, 15, 68, 110, 114, 117, 184, 188
Adorno, Theodor W.: and Benjamin,
216n95; and critical theory
development, 2, 6–10, 19–20, 25–64,
140, 188–189, 193, 207n32; on cultural
critique, 183; on empiricism and logic,
49–60; on Enlightenment, 170;
Jewishness of, 206n19; and Kracauer,
156, 182, 233n3; on lament, 95–96; on
language and philosophy, 29–40, 145;
on logical positivism, 190, 191–192,
196, 236n31; on materiality, 156; on
mathematics and Fascism, 188; on
mathematization of nature, 206n20; on
messianism, 107; on protocol
sentences, 40–49; on Rosenzweig,
231n56; and Russell's paradox, 213n35;
on subjectivity, 29–40, 116; works:
"The Actuality of Philosophy," 34,
212n34, 225n5; *Aesthetic Theory*, 148;
"The Curious Realist," 233n3; *Dialectic
of Enlightenment*, 26, 37–38, 40, 47, 48,
50, 54, 56–60, 173, 184, 189, 193,
238n63; "The Essay as Form," 38;
Minima Moralia, 42, 44, 226n8
aesthetics: and critical theory, 9, 16, 18;
and digital humanities, 198; Kracauer
on, 22, 148, 164–165, 169; and logical
positivism, 28, 64; mediation of, 234n9;
and negative mathematics, 188–189; of

privation, 16; Scholem on, 2, 67–68, 82,
98; of theory, 173–182; transcendental,
234n10, 237n52. *See also* negative
aesthetics
Albert, Hans, 62
algebra, 67, 77, 89, 132, 134
allegory, 33
Alliance of Digital Humanities
Organizations, 242n20
allusions, 83–84
Altmann, Alexander, 12
analogies, 16, 83–88, 91–92, 103, 105,
118–119, 221n46
anti-Semitism, 37, 51, 120
Aquinas, Thomas, 121, 123, 229n44,
230n46, 240n4
Aristotle, 61, 115, 121, 122, 230n49,
235n18
assimilation, 7, 67–68, 70, 76, 89, 100,
102–103, 193
authoritarianism, 5, 20, 27, 45, 53, 58, 59
axioms, 52, 110, 122, 149, 151–154, 164,
175
Ayer, A. J., 210n9

Bacon, Francis, 5, 6, 26, 43, 56, 205n8
barbarism, 7–8, 11, 15, 26, 47–48, 58,
64–66, 105, 168
Batnitzky, Leora, 127
Bauch, Bruno, 88, 210n10, 222n65

Baumeister, Karl August, 227*n*21
Bendavid, Lazarus, 208*n*37
Benjamin, Walter: on language and
    mathematics, 9, 19, 27, 28, 30, 32–33,
    34, 35, 63, 207*n*32, 219*n*22; on Poe,
    239*n*80; and Rosenzweig, 106, 139; and
    Russell's paradox, 213*n*35; and
    Scholem, 68–69, 73–74, 81–82, 98, 100,
    211–212*n*25, 216*n*95, 223*n*71, 224*n*79;
    works: "On Language as Such and on
    the Language of Man," 32, 73, 74, 82,
    92; *The Origin of German Tragic Drama*,
    33, 82, 212*n*33; "The Task of the
    Translator," 224*n*79
Bergson, Henri, 120
Berry, David, 196, 242*n*30
biblical citations and analysis, 88–99,
    220*n*36
Blumenberg, Hans, 85, 226*n*10; *Paradigms
    for a Metaphorology*, 16
Boole, George, 89, 90
Brunkhorst, Hauke, 211*n*16
Buber, Martin, 68, 87, 171, 223*n*71,
    238*n*65
Buck-Morss, Susan, 9, 234*n*9; *Origin of
    Negative Dialectics*, 207*n*26
Busa, Roberto, 240*n*4

calculus. *See* infinitesimal calculus
Cantor, Georg, 73
capitalism: and digital humanities, 198;
    Kracauer on, 22, 147, 157, 166–172,
    177–179, 185; and logical positivism,
    51–52, 60;
Carnap, Rudolf, 25, 27, 30, 41, 51, 142,
    210*n*10–11, 214*n*49; "Truth and
    Verification," 232*n*77
Cesàro, Ernest: *Lectures on an Intrinsic
    Geometry*, 174, 238*n*69
Chamberlain, Houston Stewart, 120,
    228*n*23, 229*n*41, 231*n*60
Christianity, 106, 117–119, 125, 127,
    137–138, 147, 161
Chun, Wendy Hui Kyong, 198

Cohen, Hermann: Horkheimer on,
    232*n*75; on infinitesimal calculus,
    221*n*47, 227*n*20; mathematics in work
    of, 13–14; on pure thought, 70, 83,
    221*n*45; and Rosenzweig, 105, 127–130,
    132, 139, 231*n*60; works: *The Logic of
    Pure Knowledge*, 14, 83, 86, 128; *The
    Principle of the Infinitesimal Method and
    its History*, 13; *Religion of Reason Out of
    the Sources of Judaism*, 14; *System of
    Philosophy*, 13
computational approaches, 22–23,
    190–193, 198
consciousness, 7, 35, 53, 155, 178
continuity: historical, 27, 29, 41, 96, 98,
    101, 140, 192; Scholem on, 21, 68–69,
    102–103. *See also* discontinuity
coordinate system, 160, 174, 178
Copernican Revolution, 77, 176
Cornelius, Hans, 210*n*10
Couturat, Louis, 222*n*62
criticality, 173, 193–194
critical theory: afterlife of mathematics
    in, 60–64; in digital age, 187–199;
    histories of, 9, 148; and logical
    positivism, 25–64; mathematics in,
    1–23; origins of, 2, 17, 25, 59, 61, 196.
    See also Frankfurt School
cultural critique: and critical theory,
    16–17; and digital humanities, 194,
    197; and geometry, 182–185; and
    infinitesimal calculus, 106, 109, 139,
    144; and negative aesthetics, 67, 100,
    103; and negative mathematics, 187. *See
    also* mass culture
cultural theory, 103, 140, 144, 158,
    180–183, 188

Dahms, Hans-Joachim, 51, 209*n*2, 210*n*9,
    215*n*69
Dathe, Uwe, 220*n*44
Dedekind, Richard, 84, 221*n*50
democratization of knowledge, 196
deprivation, 21, 68, 76, 89, 93

Derrida, Jacques, 106–107
Descartes, René, 149, 174–175, 178,
    239*n*77; *Optics*, 174, 238*n*70
detective novels, 147, 157–166
determinate negation, 37, 213*n*40
differentials, 12–13, 107, 118–119, 124,
    129–130, 136
differentiation, 102, 107, 114–115, 119,
    121, 124–125, 129
digital humanities, 1–2, 10, 19, 22–23,
    187, 190–199, 240*n*6, 241*n*11, 242*n*24
Dimock, Wai Chee, 144
discontinuity, 10, 16, 19, 68–69, 100, 103,
    183–184, 190, 192, 198
Du Bois, W. E. B., 242*n*32
Düttmann, Alexander, 60
dynamism, 118, 120–121, 134

Ehrenberg, Rudolf, 117
empiricism, 28, 49–51, 53, 56, 58, 62
Engel, Amir, 217*n*11, 223*n*71
Enlightenment: and critical theory, 2–12,
    14, 17, 22; and geometric space,
    146–148, 180, 182, 184; and geometry
    of modernity, 166–173; and logical
    positivism, 26, 29, 37, 43, 47, 49–50,
    56–59, 62; and negative mathematics,
    188; and reason, 22, 170
epistemology: and critical theory, 2, 21;
    and digital humanities, 195, 196; and
    infinitesimal calculus, 109, 116,
    122–123, 140; and logical positivism,
    51, 54, 63; and negative aesthetics, 75,
    84; and negative mathematics, 191
Euclidean geometry, 92, 110–113,
    121–123, 149–150, 152, 168–169
experience: and critical theory, 20–21, 27,
    29–30, 33–34, 36, 55, 57, 60–61; and
    geometric space, 146–147, 149–153,
    156–157, 164, 176, 178, 181; and
    infinitesimal calculus, 105, 109,
    112–113, 115–116, 126, 128, 136, 138,
    142; and language, 4; of modernity,
    15–18; and negative aesthetics, 76–78,

83, 87, 90, 93, 99; and negative
    mathematics, 188, 193; and reason,
    12–13; and subjectivity, 40–49
expression, 38, 45, 68, 78, 83, 95–96, 98,
    135
extremity, 8, 93–95, 99

Fascism, 22, 27–28, 40–49, 50, 53, 188,
    210*n*9
Feenberg, Andrew, 9, 49, 211*n*16, 214*n*64;
    Philosophy of Praxis, 207*n*26
Fenves, Peter, 32, 68, 207*n*32, 219*n*25
Fichte, J. G., 228*n*25
Foucault, Michel, 100, 224*n*83
fractal geometry, 144
Frankfurt School, 25–28, 33, 45, 62, 126,
    157, 206*n*19, 236*n*33
Frege, Gottlob, 29, 70, 79, 82, 89, 90
Freud, Sigmund, 236*n*35
Frisby, David, 237*n*43
Funkenstein, Amos, 208*n*46

Galileo, 5
Galison, Peter, 219*n*26
generative negativity of mathematics, 86,
    107, 110, 190
geometry, 145–185; and aesthetics of
    theory, 147, 161, 173–182; and axioms,
    150–157; and critical theory, 2, 5, 8; and
    cultural critique, 182–185; descriptive,
    149, 157–166; fractal, 144; and Husserl,
    149, 150, 152–154, 235*n*23; and pure
    sociology, 150–157; and infinitesimal
    calculus, 110–116; and Kant, 176; and
    language, 145–146, 156, 174; and
    Lukács, 147, 169, 171; and materiality,
    146–148, 150–151, 153, 156–157, 160,
    164–166, 168, 176, 180, 182–183, 197;
    and material logic, 18, 182–185; and
    metaphysics, 147, 156, 159, 165, 170; of
    modernity, 15–18, 166–173; and
    motion, 156; natural, 173–182; and
    negative aesthetics, 67, 77; and
    projection, 22, 146–147, 157–166, 169,

geometry (cont.)
171–174, 177, 181; and reality, 156, 159, 182; and representation, 159, 162. *See also* Euclidean geometry

German Sociology Society, 61

Glatzer, Nahum Norbert, 218*n*18

Gnosticism, 221*n*48

God: attributes of, 14; and infinitesimal calculus, 106, 117; language of, 32, 73–74, 102; and mathematics, 82, 85, 87–88; and messianic theory of knowledge, 9, 21, 126–127, 129–136; and metaphysics, 79; and mysticism, 93, 98

Gödel, Kurt, 84–85

Goethe, Johann Wolfgang, 114

Goldberg, Oskar, 87, 220*n*36

Golumbia, David, 193, 194, 195, 242*n*27, 242*n*30

Google Books, 191, 241*n*9

Gordon, Peter, 208*n*46, 210*n*10

grammar, 73, 133, 136

Guyer, Paul, 231*n*62

Habermas, Jürgen, 8, 58, 60, 62–63, 184, 188, 197; *Dialectic of Enlightenment*, 207*n*26; *Knowledge and Human Interest*, 62; *Theory of Communicative Action*, 216*n*94

Hahn, Hans, 4, 47

Hansen, Miriam, 147, 176, 236*n*29

Hegel, G. W. F., 28, 51, 106, 109, 131

Heidegger, Martin, 33–34

Hilbert, David, 70, 79, 152, 174, 227*n*15

historicism, 4, 72, 205*n*9; crisis of, 4, 72, 205*n*9

history, metaphysics of, 120, 147, 165, 169, 174, 176

Hofmannsthal, Hugo von, 4, 72, 205*n*8

Hohendahl, Peter Uwe, 243*n*37

Hölderlin, Friedrich, 224*n*79

Hollander, Dana, 225*n*4

Honneth, Axel, 62

Horkheimer, Max: and critical theory development, 2, 6–11, 19–20, 25–64, 140, 188–189, 193, 207*n*32; on empiricism and logic, 49–60; on Enlightenment, 170; on Kracauer, 182; on language and philosophy, 29–40, 145; on logical positivism, 190, 191–192, 196, 214*n*58, 214*n*62; on materiality, 156; on mathematics and Fascism, 188; on mathematization of nature, 206*n*20; on protocol sentences, 40–49; on subjectivity, 29–40, 116; works: *Dialectic of Enlightenment*, 26, 37–38, 40, 47, 48, 50, 54, 56–60, 173, 184, 189, 193, 238n63; *The Eclipse of Reason*, 47, 48, 173; "The Latest Attack on Metaphysics," 27, 31, 36, 45–46, 47, 49–50, 53, 55, 58–59, 193; "Traditional and Critical Theory," 189

humanism, 1–2, 9, 23, 110, 179, 189–190, 198, 240*n*3

Hume, David, 76

Husserl, Edmund: and critical theory, 3–9; and geometry, 149, 150, 152–154, 235*n*23; on mathematization of nature, 206*n*20; phenomenology of, 207*n*32, 235*n*17; works: *Cartesian Meditations*, 3; *Ideas: General Introduction to Pure Phenomenology*, 152; "The Crisis of the European Sciences and Transcendental Phenomenology," 3, 205*n*7

Huyssen, Andreas, 177, 239*n*83

idealism, 106, 109, 113, 129

Idel, Moshe, 221*n*48

immediacy, 42–43, 53, 55, 57, 76, 193

infinite judgment, 128, 221*n*47, 231*n*62

infinitesimal calculus, 104–144; Cohen on, 221*n*47, 227*n*20; and cultural critique, 106, 109, 139, 144; and epistemology, 109, 116, 122–123, 140; and experience, 105, 109, 112–113, 115–116, 126, 128, 136, 138, 142; and geometry, 110–116; and language, 105–106, 111, 114–115, 131–136, 140, 143; and logic, 121, 128, 134, 138; and messianic theory of knowledge, 140–144; and metaphorics of motion, 106, 117–126; and metaphorics of subjectivity, 106, 109–117; and motion, 104–108, 112–115, 117–130, 132, 134–136, 139, 141, 144; and negative mathematics, 194; and reality, 109, 130, 133; and representation, 126–139, 140, 142, 144; and silence, 136–137; and subjectivity, 109–117; and time, 117–126

infinity, 80, 108, 124–125, 131, 138–139

Institute for Social Research, 27

instrumental reason, 20, 26, 29, 47–49, 58–60, 64, 166, 188, 197

International Congress for the Unity of Science, 27, 30, 40, 61

Islam, 126, 230*n*55

Jay, Martin, 9, 234*n*9; *Dialectical Imagination*, 207*n*26

Jews and Judaism: and aesthetics of theory, 147, 161; and anti-Semitism, 37, 51, 120; and critical theory, 14, 18; and Frankfurt School, 206*n*19; in interwar period, 207*n*29; Kabbalah, 69, 87, 100–101, 219*n*21, 220*n*36, 221*n*48; and modernity, 216*n*5; mysticism, 3, 68–69, 79, 83–88, 93, 100–101, 103; and negative aesthetics, 68; persecution of, 101; and Rosenzweig's messianism, 106, 115, 117–119, 122, 125, 127, 137–139, 143; Scholem's analysis of Jewish lament, 88–103

Jockers, Matthew, 192, 198, 241*n*9, 241*n*15

Kabbalah, 69, 87, 100–101, 219*n*21, 220*n*36, 221*n*48

Kafka, Franz, 207*n*29, 240*n*88

Kant, Immanuel: and critical theory, 12; and idealism, 106, 109, 113, 120; on infinite judgment, 231*n*62; and natural geometry, 176; and negative aesthetics, 73, 76–77, 81, 83; on Newton, 124; on rationalism and reason, 30, 51, 52, 70, 76; on theory of knowledge, 128; on transcendental aesthetic, 234*n*10, 237*n*52; and visual-geometric frame of knowledge, 92; works: *Critique of Pure Reason*, 30, 70, 76

Kierkegaard, Søren, 161, 162

Kirsch, Adam, 191, 192, 241*n*11

Klein, Felix, 110, 111, 116, 227*n*19

Klein, Lauren, 19, 193

knowledge: democratization of, 196; messianic theory of, 140–144; modes of, 45, 84; positive, 14, 131; sociological, 150; theory of, 116, 123, 160, 188; totality of, 50, 106; true, 33, 55

Koch, Gertrud, 238*n*62

Kowalewski, Gerhard, 229*n*34

Kracauer, Siegfried, 21–22, 145–185; on aesthetics of theory, 173–182; on axioms of necessity, 150–157; and critical theory, 2–3, 8–11, 15–19; on cultural critique, 182–185, 197; on descriptive geometry, 149, 157–166; on geometry of modernity, 166–173; on pure sociology, 150–157; on material logic, 182–185; on natural geometry, 173–182; and negative mathematics, 189, 190; on relativism, 206*n*9; on Rosenzweig, 126, 228*n*24; works: "Analysis of a City Map," 173, 177; "The Bay," 178–179, 180; "The Crisis of Science," 72; *The Detective Novel: An Interpretation*, 157–166, 169, 173; *Ginster*, 159; *History: The Last Things before the Last*, 184–185; "Lad and Bull," 173; "Lead In: Natural Geometry," 174, 176, 180; "The Mass Ornament," 159, 166–173, 179, 180–181, 197, 238*n*63; "Photography," 159; *Sociology as Science*, 150–157, 161, 173; "Two Planes," 173, 177

Kraft, Werner, 219*n*22

Kraus, Karl, 181
Krebs, Engelbert, 229*n*44

lament and lamentations, 20–21, 66–67,
    69, 88–103, 180, 192
language: analogies, 16, 83–88, 91–92,
    103, 105, 118–119, 221*n*46; audible, 134;
    and critical theory, 4, 8–11, 15, 19–21,
    23; divine, 32, 73–74, 102; formalized,
    41, 90; and geometry, 145–146, 156,
    174; grammar, 73, 133, 136; historical,
    27, 29, 41; and infinitesimal calculus,
    105–106, 111, 114–115, 131–136, 140,
    143; of Jewish lament, 88–99; limits of,
    82–88; literary, 94–95; and logical
    positivism, 26–28, 29–40, 41–42,
    44–46, 49, 51, 53, 55, 60;
    mathematization of, 42, 134; and
    negative aesthetics, 65–70, 72–75, 78,
    80–96, 98–101, 103; and negative
    mathematics, 192, 198; phonetic, 65,
    90–91, 93; poetic, 8, 68, 89, 94–96, 99;
    primordial, 131–132, 134–135; protocol
    sentences, 40–49; structure of, 70–75;
    theological, 105, 118, 142, 156. *See also*
    metaphorics
Lask, Emil, 211*n*16
Lazier, Benjamin, 73
Leibniz, Gottfried Wilhelm, 4–5, 12, 21,
    51, 79, 107, 114, 119, 128, 129
liturgy, 131, 135, 136–138
Liu, Alan, 187, 193–194
logic: and critical theory, 13, 18, 28–31,
    34–37, 39, 42, 44, 49, 51, 53, 55–56,
    61–62, 65; dialectical, 61; in geometry,
    18, 147, 151, 154–158, 160, 164–170,
    172, 176, 180, 183; and infinitesimal
    calculus, 121, 128, 134, 138; and
    negative aesthetics, 70, 73, 80, 88–92;
    and negative mathematics, 197–198;
    and Scholem's analysis of Jewish
    lament, 88–99
logical positivism, 25–64; Adorno on, 190,
    191–192, 196, 236*n*31; and aesthetics,

28, 64; and Benjamin, 19, 27, 28, 30,
    32–33, 34, 35, 63; and capitalism, 51–52,
    60; and critical theory, 7, 19–20, 23,
    25–64, 190–191, 193, 214*n*62; critiques
    of, 26, 40; and empiricism, 49–60; and
    Enlightenment, 26, 29, 37, 43, 47,
    49–50, 56–59, 62; and epistemology, 51,
    54, 63; and geometry, 145, 182;
    Horkheimer on, 190, 191–192, 196,
    214*n*58, 214*n*62; and language, 26–28,
    29–40, 42–46, 49, 51, 53, 55, 60; and
    Lukács, 52–53, 54, 55; and materiality,
    30, 34–35, 63; and metaphysics, 26–31,
    34, 36, 41, 45–47, 49–60; and Nazism,
    214*n*58; and negative mathematics, 196;
    and protocol sentences, 40–49; and
    rationalism, 49–64; and reality, 36, 39,
    44, 51–55, 61; return of, 56–57; and
    silence, 31, 44–46
Lotze, Hermann, 89, 90, 191; *Logik*, 88
Löwenthal, Leo, 126, 159
Lukács, George: and critical theory, 9,
    19; and geometry, 147, 169, 171; and
    logical positivism, 52–53, 54, 55; on
    materiality, 156; works: *History and
    Class Consciousness*, 167, 211*n*16, 213*n*36,
    215*n*75, 236*n*29; *Theory of the Novel*, 150,
    156, 167, 233*n*6

Mach, Ernst, 213*n*46
Maimon, Salomon, 12–13, 208*n*37,
    208*n*39; *Essay on Transcendental
    Philosophy*, 12
Maimonides, 14
Marburg School, 231*n*60
Marcuse, Herbert, 8, 215*n*87; *One-
    Dimensional Man*, 61
Marx, Karl, 28, 56, 166, 237*n*56
mass culture: and aesthetics of theory,
    148, 157, 159, 164, 172–173, 185; and
    critical theory, 8, 11, 22. *See also*
    cultural critique
materiality: and critical theory, 8, 15, 18,
    22; of experience, 151–153; and

geometry, 146–148, 150–151, 153, 156–157, 160, 164–166, 168, 176, 180, 182–183, 197; and logical positivism, 30, 34–35, 63

material logic, 8, 18, 182–185

mathematics: in critical theory, 1–23; generative negativity of, 86, 107, 110, 190; and education, 109–117; and language, 29–40, 73, 80, 82, 135; logic and reasoning in, 2, 18, 20, 29–35, 37, 39, 42–43, 46, 53, 57, 61, 65–66, 69, 76–82, 86, 88–99, 102, 106, 113–114, 129, 140, 154, 156, 188, 191–193; and mysticism, 83–85, 87, 100; philosophy of, 65–103; privative structure of, 75–82, 102–103. *See also* geometry; infinitesimal calculus

mathematization: of language, 42, 134; of literature, 198; of logic, 44; of nature, 5–6, 206*n*20; of thought, 2, 7–8, 59, 182, 191

Mauthner, Fritz, 4, 72, 73, 77, 80, 90; *Contributions to a Critique of Language*, 72–73

Mehrtens, Herbert, 217*n*12, 227*n*15

Mendelssohn, Moses, 228*n*23; "On Evidence in the Metaphysical Sciences," 11–12

Meraner Reforms, 110

messianism: and critical theory, 8, 17, 21; and mathematics, 187–188; and negative aesthetics, 64; of Rosenzweig, 105–107, 109, 117, 122, 126–127, 135–140, 142; and theory of knowledge, 9, 105, 107, 109, 140–144, 184

metaphorics: of construction, 219*n*26; of geometry, 153, 155–156; of infinitesimal calculus, 106, 109–126, 132–135; of motion, 105–106, 108, 112, 117–127, 130, 132, 139, 156; of projection, 236*n*35; of space, 22, 146, 148, 150, 153–154, 156, 158, 160, 169, 172–174, 176, 180–182; of structure, 66, 69–70, 72, 75, 77, 82–83, 85, 88; of subjectivity,

105–106, 108, 109–117, 126–127, 130, 132, 139, 156

metaphysics: and critical theory, 11–12, 14; and geometry, 147, 156, 159, 165, 170; of history, 120, 147, 165, 169, 174, 176; and logical positivism, 26–31, 34, 36, 41, 45–47, 49–60; and mathematics, 13–14; meaning in, 147, 161, 163–164

Minimal Computing, 194, 242*n*20, 243*n*34

modernity: and critical theory, 2, 7–8, 15–19, 29, 56; geometry of, 146–147, 157, 165, 166–173, 175–177, 180–181, 183

Molitor, Franz Joseph, 222*n*58

Monge, Gaspard, 158

Moretti, Franco, 191, 192, 241*n*14

motion: continuous, 119, 123–124, 129, 138; and critical theory, 12, 20, 63; and geometry, 156; and infinitesimal calculus, 104–108, 112–115, 117–130, 132, 134–136, 139, 141, 144; metaphorics of, 106, 117–118, 121–123, 129–130, 138, 139, 141

mourning, 94–96

Mülder-Bach, Inka, 155, 236*n*41, 237*n*55

mysticism, 3, 68–69, 79, 83–88, 93, 100–101, 103

myth, 15, 28, 38, 49, 56–62, 66, 82, 89, 105, 170

natural geometry, 22, 149, 173–176, 180–181

National Socialism, 3–4, 27, 45, 51, 58, 205*n*5, 214*n*58

negation, 14, 21, 37, 67, 84, 130, 132, 135, 213*n*40

negative aesthetics, 65–103; and critical theory, 8; and cultural critique, 67, 100, 103; and epistemology, 75, 84; and experience, 76–78, 83, 87, 90, 93, 99; and geometry, 67, 77; as history and tradition, 99–103; and infinitesimal calculus, 113, 140; and language,

negative aesthetics (cont.)
    65–70, 72–75, 78, 80–96, 98–101, 103;
    and logic, 70, 73, 80, 88–92; and
    logical positivism, 75–76, 99; and
    mathematical Platonism, 82–88; and
    metaphorics of structure, 70–75; and
    metaphysics, 99; and modernity, 66;
    and privation, 65, 67–69, 75–82, 84–85,
    88, 94, 98, 100–103; and reality, 71, 73,
    80, 88; and representation, 65–66, 75,
    82–85, 87, 89, 91, 95, 99, 103; and
    silence, 67–69, 89, 91–96, 98–101, 103,
    223*n*71
negative mathematics, 14–21; and critical
    theory, 10–14, 23, 39, 49, 187–199; and
    cultural critique, 182–183; and digital
    humanities, 23, 188–198; and
    geometry, 148–150, 156, 158, 165–166,
    168, 171, 173, 180–185; and
    infinitesimal calculus, 107, 109, 113,
    115, 128, 132, 136–142; and metaphors,
    15–19; and negative aesthetics, 67, 76,
    89, 99–103; project of, 69–70, 115, 140,
    187, 197; relevance of, 18, 185
Neo-Kantians, 77, 83, 128
neo-positivism, 61
Nernst, Walter, 218*n*16; *Introduction to the
    Mathematical Treatment of the Natural
    Sciences*, 71
Neurath, Otto, 25, 27, 210*n*6, 210*n*9,
    214*n*49; *Empiricism and Sociology*,
    210*n*8
Newton, Isaac, 12, 21, 107, 114, 121–123,
    128–129, 229*n*37; *On the Quadrature of
    Curves*, 119, 123; *Philosophiæ Naturalis
    Principia Mathematica*, 122
Nietzsche, Friedrich, 72, 77, 207*n*26
nihilism, 72–73, 75, 77, 92
North, Paul, 207*n*29
nothingness, 105, 108, 129–131, 136
Novalis, 87, 220*n*33, 227*n*20

Omeka (digital humanities platform),
    194, 242*n*20

Peano, Giuseppe, 89
pedagogy, 21, 70, 109–110, 116, 174, 194
perception, 5, 37, 42–43, 86, 148, 164,
    174, 176
phenomenology, 3, 150, 152–155, 207*n*32
Pines, Shlomo, 230*n*55
Piper, Andrew, 241*n*14
Plato, 51, 70, 82–88, 92, 221*n*48; *Meno and
    Other Dialogues*, 228*n*23; *Theaetetus*,
    228*n*23
Poe, Edgar Allen, 178, 239*n*80
poetics, 21, 70–71, 76, 96, 99, 103, 188
poetry, 28, 30, 37–38, 41, 99, 140, 192
Poincaré, Henri, 70, 75–76, 78
Pollock, Benjamin, 21, 106, 129, 218*n*18,
    226*n*7, 228*n*31
Popper, Karl, 62, 232*n*77
positivism, 56–57, 62–63, 188. *See also*
    logical positivism
Posner, Miriam, 197
primordial language, 131–132, 134–135
primordial phenomenon, 113–114, 117,
    120
Pringsheim, Alfred, 227*n*19
privation: and critical theory, 8, 10, 15–17,
    20–21, 188; historical, 96, 98, 192; and
    Jewish lament, 94; of language, 95–96
projection: and geometry, 22, 146–147,
    157–166, 169, 171–174, 177, 181; as
    metaphor, 236*n*35; method of, 22, 146,
    161, 182; parallel, 237*n*46; of
    rationality, 174, 178–180
protocol sentences, 38, 40–49, 214*n*49
pure knowledge, 13–14, 83, 86, 128

Rabinbach, Anson, 11, 107, 207*n*29
rationality and reason: autonomy of, 178;
    and critical theory, 2–7, 9–10, 12, 14,
    17, 19–20, 22; and geometry, 146–149,
    156, 159, 165–174, 177–182, 184–185;
    and geometry of modernity, 166–173;
    and infinitesimal calculus, 104;
    instrumental reason, 20, 26, 29, 47–49,
    58–60, 64, 166, 188, 197; limits of,

207*n*26; and logical positivism, 27–29, 47–48, 49–64; and negative aesthetics, 76; and negative mathematics, 188, 193, 195, 197–198; of space, 180

rational numbers, 125, 138

reality: and critical theory, 8; and geometry, 156, 159, 182; and infinitesimal calculus, 109, 130, 133; and logical positivism, 36, 39, 44, 51–55, 61; non-sensual, 84–85; self-assured, 125

reason. *See* rationality and reason

redemption, 11, 15, 18, 93, 106–108, 126, 137, 142

Reichenbach, Hans, 232*n*77

relativism, 72, 153, 157, 205*n*9

representation: and geometry, 159, 162; and infinitesimal calculus, 126–139, 140, 142, 144; in language, 32–37, 40; and logical positivism, 51; modes of, 36, 92; and negative mathematics, 194

Riecke, Eduard, 226*n*12

Risam, Roopika, 193, 194

Rosenstock, Eugen, 106, 117, 118, 228*n*31

Rosenzweig, Franz, 104–144; and critical theory, 2–3, 8–11, 15–19, 21, 140–144; Glatzer on, 218*n*18; and Jewish identity, 68; Kracauer on, 171; messianic theory of knowledge of, 140–144; on metaphorics of analogy, 221*n*46; on metaphorics of motion, 117–126; on metaphorics of subjectivity, 109–117; and negative mathematics, 189, 190, 194, 195; on representation, 126–139; on time, 117–126; works: "The New Thinking," 140–142; *The Star of Redemption*, 9, 21, 105, 109, 111, 125–140, 198, 208*n*46, 228*n*24; "*Volksschule* and *Reichsschule*," 109–114, 115, 118, 120–121, 129, 137, 228*n*23

Russell, Bertrand, 29, 33, 55, 82, 89, 90, 212*n*27

Russell's paradox, 32, 212*n*25, 212*n*27, 213*n*35

Salkowski, Erich, 234–235*n*14

Santner, Eric, 135

Scheebart, Paul: *Lesabéndio*, 92

Schegel, August Wilhelm, 214*n*57

Schelling, F. W. J. von, 109

Schiller, Friedrich, 171

Schlegel, Friedrich, 220*n*33

Scholem, Gershom, 65–103; and Benjamin, 211–212*n*25, 216*n*95, 224*n*79; and critical theory, 2–3, 8–11, 15–19, 20–21; father's disapproval of, 217*n*6; on Jewish lament, 88–103; on language and mathematics, 219*n*22; on metaphorics of structure, 70–75; on modernity and Jewish families, 216*n*5; and negative mathematics, 189, 190, 193; on privative structure of mathematics, 75–82, 102–103; on Rosenzweig's messianism, 139, 140; works: *Major Trends in Jewish Mysticism*, 21, 68, 101, 192, 198; "A Medieval Lamentation," 96; "On Lament and Lamentation," 93, 94, 96; *Referat* on mathematical logic, 88, 90–93, 96, 212*n*25, 222*n*65

Schönflies, Arthur, 218*n*16; *Introduction to the Mathematical Treatment of the Natural Sciences*, 71

Schopenhauer, Arthur, 111

Schreber, Daniel Paul, 236*n*35

Schubring, Gert, 227*n*17

self-preservation, 26, 48, 58

silence: and aesthetics of theory, 163; and critical theory, 4, 19–21; and infinitesimal calculus, 136–137; language of, 92–93, 95; and logical positivism, 31, 44–46; and negative mathematics, 187, 190, 192–193, 198

Simmel, Georg, 152, 153, 161, 165, 235*n*20, 236*n*41; "On the Spatial Projection of Social Forms," 160; *The Philosophy of Money*, 152

skepticism, 3, 20, 71–78, 92, 145, 153

Smith, John H., 115

space: and critical theory, 8, 15, 17, 22; and geometry, 145–146, 148–150, 169, 172–174, 176, 178–183; and infinitesimal calculus, 113, 121, 123–124, 128, 141; liminal, 183; metaphors of, 146, 153–155, 171–172; and negative aesthetics, 77, 83; and negative mathematics, 188; and projection, 157–166; rationalization of, 180–181; textual, 174, 180; three-dimensional, 149, 174, 177

spheres, 32, 51, 65, 73, 155, 160–161, 163

Spinoza, Baruch, 154

Steiner, Rudolf, 87

STEM programs, 190, 240*n*3

subjectivity: dynamic, 135–136; and infinitesimal calculus, 109–117; metaphorics of, 109–117, 121, 136–137, 139; removal of the subject, 38, 40–49

Susman, Margarete, 112, 145, 157

symbols: and critical theory, 10, 20; and infinitesimal calculus, 112, 133; and logical positivism, 26, 28–29, 32, 35, 39, 51, 54, 61, 63; and negative aesthetics, 68, 81, 83–84, 89–92, 96, 98, 101, 103–104; and negative mathematics, 188, 192

tautologies, 30–31, 34, 55–58, 75–80

Taylorism, 149

temporality, 117–119

Tieck, Ludwig, 220*n*33

Tiedemann, Rolf, 212*n*25

Tiller Girls, 147, 166, 168, 169–170, 172, 181

totalitarianism, 45, 47

Troeltsch, Ernst, 4, 72; "The Crisis of Historicism," 205*n*9

truths: and critical theory, 12, 16, 21; eternal, 142; and geometry, 147, 155, 168, 170, 172; and infinitesimal calculus, 105, 109, 112, 138, 141–142; and logical positivism, 31, 33–34, 36, 43–44, 56–58; and negative aesthetics, 69, 72–73, 79, 85, 102–103; and negative mathematics, 188; theological, 105, 142

unity, 58, 85, 151, 156

Vidler, Anthony, 163

Vienna Circle: and critical theory, 19; and empiricism, 210*n*8; and logical positivism, 25, 27–28, 33, 35, 41, 45, 47, 50, 196; works: *The Scientific Conception of the World*, 28, 41, 50, 55, 211*n*24

violence, 2, 7, 11, 58–59, 62

Volkert, Klaus, 70, 92, 206*n*11

von Moltke, Johannes, 172, 179, 233*n*4, 239*n*82

Voss, Aurel, 84; *On the Essence of Mathematics*, 67, 71

Weatherby, Leif, 225*n*1, 227*n*20

Weber, Max, 5, 29, 66, 72, 206*n*9

Weber, Samuel, 69

Weierstraß, Karl, 102, 227*n*15

Weierstraß function, 206*n*11, 225*n*88

Wellmer, Albrecht, 59–60

Wheatland, Thomas, 211*n*15

Whitehead, Alfred North, 55, 82, 89, 90

Wiggershaus, Rolf, 213*n*39

Wittgenstein, Ludwig, 4, 30, 37, 43, 213*n*46

World War I, 106, 111

World War II, 47–48

Zeno's paradoxes, 123

Printed and bound by CPI Group (UK) Ltd, Croydon, CR0 4YY

27/10/2024

14580327-0003